Activities and Proje
for

INTRODUCTORY STATISTICS COURSES
Second Edition

RON MILLARD
Shawnee Mission South High School
Graceland University

W. H. Freeman and Company
New York

© 2008 by W. H. Freeman and Company

ISBN-13: 978-0-7167-6544-8
ISBN-10: 0-7167-6544-6

All rights reserved.

Printed in the United States of America

First printing

W. H. Freeman and Company
41 Madison Avenue
New York, NY 10010
Houndmills, Basingstoke RG21 6XS, England

www.whfreeman.com

Contents

Preface		v
Notes to Instructors		vii

PART I

Project 1	Spring Break	1
Project 2	Population Growth	5
Project 3	Name Brand vs. Off Brand	7
Project 4	The Age of a Penny	9
Project 5	Snap Crackle Pop	12
Project 6	Movin' Out	16
Project 7	Short or Tall	18
Project 8	Skittles Machines	21
Project 9	NBA	26

PART II

Project 10	What's the Cost of Lunch?	28
Project 11	What's It Worth?	31
Project 12	Test Your Memory	33
Project 13	What Do Students Drive?	36
Project 14	Jellybean	38
Project 15	Fair Coin	40
Project 16	Odd or Even	43
Project 17	Ahoy Mates	46
Project 18	Taste the Difference	49
Project 19	Plain and Peanut	53
Project 20	Many Variables Do Predict	59
Project 21	Growth Groups	61
Project 22	Random Assignment	64
Research Project	Project Final	66

STATISTICS IN RESEARCH

Preface for Statistics in Research	69
Sarcopenia, Calf Circumference, and Physical Function of Elderly Women	77
Opinion Formation in Evaluating Sanity at the Time of Offense	87
Imagination, Personality, and Imaginary Companions	105
Influence of Seeding Depth and Seedbed Preparation on Establishment, Growth and Yield of Fiber Flax in Eastern Canada	125
Shape of Glass and Amount of Alcohol Poured	135
Linking Superiority Bias in the Interpersonal and Intergroup Domains	141
Monitoring Early Reading Development in First Grade	157
Online Issues of the *Journal of the American Medical Association*	174
Association of Long-Distance Corridor Walk Performance with Mortality, Cardiovascular Disease, Mobility Limitation and Disability	177

Preface

The statistics classroom is a learning lab where teacher and learner interact to find understanding. Statistical concepts are learned when the student is enmeshed in the process. All students do not learn from the same modalities, and my experience has shown that using hands-on projects assists learning. For example, food in the classroom captivates student attention and heightens their involvement in the lesson. When motivation is present, students will learn. Learning is a life-long journey. I convince my students that "if something makes sense today, then it will make dollars later." Statistical understanding is the model for continued improvement in the consumer products and services we use. Learning how to learn and make decisions will make today's students into the leaders of the future.

Many of the ideas for these projects have been developed through the sharing of ideas with fellow teachers. Thanks to each of them for sharing their ideas, as I likewise shared mine with them. Change comes with each new class, and as a teacher, you are challenged to meet the needs of that group of students. New understandings over time have modified the original forms of most of these activities, and with each new semester, I continue to adjust them to fit the needs of the students. The sophistication of calculators has changed most of the random experiments and provided learning of new methods for random simulations. However, I still find that rolling a die, using a spinner, and tossing coins provide the student with conceptual insights about randomization. The Internet continues to provide resources for data. Students must use these tools in a hands-on environment where they can explore and seek solutions to open-ended questions to be prepared for their future. The education and learning today assists their future ability to learn.

The projects in this supplement are divided into two parts. Part I includes projects that can be given to students to do on their own or can be used in the classroom. Part II represents those projects that are designed primarily for classroom use with teacher input and guidance. I have indicated the supplies and materials necessary for use in each project in the Notes to Instructors section. The scope and sequence of the projects is similar to the order in which topics are covered in many texts. In many of the later projects, graphs are used in addition to the hypothesis-testing formats so that students have the opportunity to show the data in addition to the thinking, statistical reasoning, and writing conclusions in plain English vs. mathematical jargon. Many science classes have a component of research that models the statistics of these projects. Cross-discipline application and learning occur as you use the skills from one class to another.

Projects and Activities can be used in a variety of ways within the class. For example, they can be used to introduce a concept, given as an application of a concept, or assigned as an enrichment activity for students.

Open-ended assessments provide students with the opportunity to discover concept application. Preparing a poster to display or making a classroom presentation are also great lessons for students to learn. Shortened class weeks due to vacation where introducing a new unit is impractical are a great learning opportunity, and Skittles or M&M's can help with the motivation for those class days.

Excitement, enthusiasm, and joy in teaching are my goals. May your students catch the fever for learning.

November 2006

Notes to Instructors

The following summary provides a statement of purpose for each project and the materials needed.

PART I

Project 1   SPRING BREAK

Purpose:
To look at the relationship of distance traveled and air fares. Students will use the Internet to find the lowest fares for a set of destinations for typical spring break trips, as well as the distance to those destinations. Possible Websites are Mapquest, Travelocity, Expedia, and individual airline Websites. A scatterplot of the data, a least-squares regression line, and correlation will be found. Students will give persuasive arguments for their conclusions.

Materials needed:
Access to the Internet

Project 2   POPULATION GROWTH

Purpose:
To use the Internet to find population data, and then use the data to make a prediction of the population for a chosen location in the next census.

Materials needed:
Access to the Internet

Project 3   NAME BRAND VS. OFF BRAND

Purpose:
To compare quality consistency of products produced by a well-known name brand to an off-brand manufacturer and to research statistical quality control methods developed by Dr. Demming.

Materials needed:
2 bags of jelly beans (solid colors) from a well-known manufacturer
2 bags of jelly beans (solid colors) from an off-brand manufacturer
Access to the Internet
TI 83/84 calculator

Project 4   THE AGE OF A PENNY

Purpose:
To determine the approximate age of pennies in circulation. The distribution of the ages, not the year the coin was minted, will be examined. The central limit theorem will be used to form an interval estimation of the mean age for each sample of coins.

Materials needed:
One roll of pennies for each student

This activity can be repeated using nickels, or any other coin, and then comparing the two distributions.

Project 5   SNAP CRACKLE POP

Purpose:
To compare two distributions using a hypothesis test for the difference of two means using the P-value approach.

Materials needed:
None. However, students will either need to go out of the classroom to the grocery store to gather their data, or they will need Internet access to search cereal company Websites.

Project 6   MOVIN' OUT

Purpose:
To compare two distributions using a hypothesis test for the difference of two means using the P-value approach.

Materials needed:
Access to the Internet

Project 7   SHORT OR TALL

Purpose:
To find a line of best fit to a paired set of data values, use inferential methods of the linear model, and make estimation and prediction intervals.

Materials needed:
None

Project 8 SKITTLES MACHINES

Purpose:
To check data with statistical process control.

Materials needed:
Fun-size bags of Skittles
TI-83/84 calculator

Project 9   NBA

Purpose:
To check data with statistical process control.

Materials needed:
Access to NBA data (Website or newspaper)
TI-83/84 calculator

PART II

Project 10   WHAT'S THE COST OF LUNCH?

Purpose:
To gather data and display it using both a graphical and a numeric method.

Materials needed:
None

Project 11 WHAT'S IT WORTH?

Purpose:
To gather data and display it using both a graphical and a numeric method.

Materials needed:
Access to the Internet

Project 12 TEST YOUR MEMORY

Purpose:
To work with residuals. Choose a list of people whose names your students recognize. www.famousbirthdays.com is a Website you can use to find birth dates of celebrities. Included here is a select group of individuals who are probably recognized nationwide.

George W. Bush	07 – 06 – 46
Catherine Zeta Jones	09 – 25 – 69
Garth Brooks	02 – 07 – 62
Whoopi Goldberg	11 – 13 – 55
LaBron James	12 – 30 – 84
Julia Roberts	10 – 28 – 67
Bill Gates	10 – 28 – 55
Queen Elizabeth II	04 – 21 – 26
Arnold Schwarzenegger	07 – 30 – 47
Brad Pitt	12 – 18 – 63
Hillary R. Clinton	10 – 26 – 47
Julie Andrews	10 – 01 – 35
Justin Timberlake	01 – 31 – 81
Dakota Fanning	02 – 23 – 94

I suggest that you use at least 10 people with a wide range in ages. Include some regional celebrities, yourself, or some well-known member of the faculty for some class fun.

Materials needed:
People known to students and their birth dates

Project 13 WHAT DO STUDENTS DRIVE?

Purpose:
To design a study and gather data.

Materials needed:
Access to a student parking lot

Project 14 JELLYBEAN

Purpose:
To use capture-recapture random sampling to estimate the number of jellybeans in a jar.

Materials needed:
1 large bag of multicolored jellybeans (with no black)
1 small bag of black jellybeans
1 large clear jar or fish bowl
Utensils for stirring and dipping (big spoons or ladles)
Paper cups or napkins

Project 15   FAIR COIN

Purpose:
To observe probability of a repeated event over time. The project may be done individually or as a group where each member of the group uses a different coin.

Materials needed:
Coins for each student

Project 16   ODD OR EVEN

Purpose:
To explore binomial probability.

Materials needed:
One die to roll
One penny to move as a game piece

An alternative procedure is to use a random number generator on your calculator in place of the die.

Project 17   AHOY MATES

Purpose:
To develop a sampling procedure, generate data, display the data with an appropriate graph, and perform a hypothesis test of a company's claims. Students are to identify procedures for designing an experiment, use a randomizing process to gather data, and conduct a hypothesis test of means.

Materials needed:
One bag of CHIPS AHOY cookies

Project 18   TASTE THE DIFFERENCE

Purpose:
To model a taste test to the binomial distribution and perform a hypothesis test for proportions. Students are to gather data and use the binomial probability density function and model the binomial distribution with the standard normal distribution. On the TI-83/84 calculator, the binomialpdf and binomialcdf functions will be used.

Materials needed:
One one-pound bag of plain M&M's (make sure the color brown is included)
TI-83/84 calculators

Project 19   PLAIN AND PEANUT

Purpose:
To analyze more than one data set and the differences in two or more distributions. The project may be used as separate parts for the comparison, or be used with the Chi Square distribution.

Materials needed:
One one-pound bag of plain M&M's
One half-pound bag of plain M&M's
One package of snack-size plain M&M's
One one-pound bag of peanut M&M's
One half-pound bag of peanut M&M's
One package of snack-size peanut M&M's
Access to the Internet
TI-83/84 calculators

Project 20   MANY VARIABLES DO PREDICT

Purpose:
To use a computer to find multiple variable equations.

Material needed:
Access to a computer with Excel or Minitab

Project 21   GROWTH GROUPS

Purpose:
To compare multiple groups using ANOVA.

Materials needed:
Access to a computer with Excel or Minitab

Project 22   RANDOM ASSIGNMENT

Purpose:
To compare multiple groups using ANOVA.

Materials needed:
Access to a computer with Excel or Minitab
Class rosters with student names

Project Final

Purpose:
To use five aspects of a statistical study. The student may use either an observational study or a designed experiment. This project is best used as the culminating activity for the class in order to demonstrate overall conceptual learning.

As the instructor, approve the appropriateness of the students' research topics before they begin their project.

Materials needed:
None

Looking at Data Relationships

Project 1   SPRING BREAK

Spring break is coming and it's time to make your plans. Below is a list of possible destinations for your trip. Fill in the following information to assist in planning your trip and finding the information needed for this project.

1. You are leaving from   _____

2. Date of departure   _____

3. Date of return   _____

You are to use the Internet to search for the lowest airfares from your town to the destination on the given dates. Also, find the distance from your airport to that city and double it to reflect round-trip mileage.

4. Complete the following table of information for your spring break adventure. The cost is for one person on a round-trip ticket. A maximum of two stops en-route are allowed, for both directions of the trip.

List the Websites that you use for your search:

Distances   _____

Airfares   _____

Destination	Round-trip Distance	Airline	Departing flight number	Round-trip cost
Orlando				
San Diego				
San Juan, Puerto Rico				
New York City				
Chicago				
Seattle				
Salt Lake City				
Boston				
Honolulu				
Denver				

5. In the space below, make a scatterplot of the distance versus cost. Use the distance as the independent variable on the horizontal axis. Make appropriate scales for the axes.

6. Look at the scatterplot above to answer the following: How is the cost of the trip associated with the distance of the trip?

7. Choose two of the data points that reflect a line of best fit to the data. Use a straight edge to draw the line. Find the slope of the line (b) and the y intercept (a). Write the equation of your estimated line:
$Y = a + bx$

8. On a scale of 0 to 1, make an estimate of how well the line fits the data. 0 = no fit and 1 = a perfect fit.

9. Using your TI-83/84 calculator, enter the data for the distances and the costs. Use L1 for the distances and L2 for the cost. Press STAT, CALC, then down arrow to LinReg($a + bx$) to find the equation of the least-squares regression line.

10. Find the value of the correlation coefficient for the data. The r value can be found as follows on the TI-83/84: Press VARS, choose Statistics, EQ, and choose r from the menu screen. As an alternative, you may go to the catalog and select DiagnosticOn to do the LinReg($a + bx$) as indicated in question 9. When DiagnosticOn is activated on your calculator, the correlation value r will automatically be displayed when you do the LinReg($a + bx$) computation.

11. Sketch the least-squares regression line on the scatterplot in Question 5 and compare it to the line that you guessed. Are they close to the same line?

12. In Part 8 you guessed a value for the strength on the fit to your line. Compare it to the value of the correlation coefficient. How close were you?

13. By looking at the scatterplot and the regression line, find a destination that is a good buy for the miles traveled. You are searching for the point that is the farthest below the line. State the destination, the expected cost by using the line of best fit, and the actual cost from the table.

   Destination   _____

   Expected cost   _____

   Actual cost   _____

14. Use the destination identified in Question 13 and any information found in this project to write a persuasive essay (50 to 100 words) convincing your parents to let you go to your destination of choice for spring break.

Looking at Data Relationships

Project 2 POPULATION GROWTH

Choose a location for which you would like to know the population. This may be a state, county, region of the country, or city. Use the Internet to search the U.S. census reports.

1. List your choice in the space provided.

   _____

2. You are to research the population given by the U.S. census reports to find the population for each of the previous census reports from 1790 through 2000. List the populations in the table below:

Year	Population
1790	
1800	
1810	
1820	
1830	
1840	
1850	
1860	
1870	
1880	
1890	
1900	
1910	
1920	
1930	
1940	
1950	
1960	
1970	
1980	
1990	
2000	

3. Use your TI-83/84 to explore models for the growth of the population over time. Make a plot of the data by choosing STAT PLOT. Use Year as the X List and Population as the Y List for the data set. In the table below, give the equation of four possible models of the data set and the value of R^2 for each model. Provide a sketch of the data and graph of the model by using the Graph Link or hand sketch for each of the four models. Attach the graphs to your report for this project.

Model	Equation	R^2

4. Use the graphs with the supporting information as listed above to make a decision on which model best fits the data.

5. Give supporting reasons to your conclusion.

6. Using your model, predict the population in the next census.

Statistics for Quality

Project 3 NAME BRAND VS. OFF BRAND

Is there a difference in the consistencies between recognizable product brands known for their quality and unrecognized brands? Does the price difference reflect more consistency of product? To look for consistency and price differences, we will use traditional jelly beans (solid colors) in the following experiment.

1. Purchase two bags of the same type of jelly beans of a very well-known brand and two identical bags of an off-brand. In the following table, list the colors used in each bag, the count of each color, and the difference between the two for each brand.

Color	Well-known bag 1	Well-known bag 2	Diff.	Color	Off-brand bag 1	Off-brand bag 2	Diff.
Total				Total			

2. Perform a paired difference test on the well-known brand.

   H_o:

   H_a:

   T = _____   P-value _____

   Word conclusion:

7

3. Perform a paired difference test on the off brand.

   H_o:

   H_a:

   T = _____ P-value _____

   Word conclusion:

4. Look at the answers to Questions 2 and 3 to see any differences in consistency between the two brands. List them.

5. Typically, well-known brands are more expensive than off-brand products. If there is a difference in the cost, is it justified in terms of the consistency of the product manufacturing?

6. Dr. William E. Demming developed statistics for quality implementation in the manufacturing process. Using the Internet, research Dr. Demming, state the company most known for using his methodology, and give an example of a result of his work.

From Probability to Inference

Project 4   THE AGE OF A PENNY

Have you ever wondered how long coins stay in circulation? Are you a collector? You each have a tube of pennies. Your first task is to form a distribution of their ages.

1. Organize the data by using a stemplot of the ages. Split the stems to give a sufficient number of stems to the data. You must have between 5 and 20 stems.

2. What is the shape of the distribution? Why do you think it is this shape?

3. Find the five-number summary for this data set.

4. Using the five-number summary and the definition of an outlier, do you find any outliers?

5. Do you think the distribution of all pennies in circulation is similar to your sample?

6. List the characteristic assumptions for the Central Limit Theorem and decide if they are satisfied by your distribution.

7. Find the mean and standard deviation of the ages of the pennies in your sample.

   Mean = _____   S.D. = _____ n = _____

8. Compute a 95% confidence interval for the mean ages of pennies.

   _____ < u < _____

9. What is the margin of error for your estimate?

   M.E. = _____

10. The president of Coins Unlimited has just hired you as his chief statistician for his research on the age of pennies. You are charged with the task of estimating the average age of pennies in circulation within one year of age with 99% confidence. Determine the sample size you would need for a one-year margin of error with 99% confidence. Use the standard deviation from your sample as your best estimate of the population standard deviation.

11. Consider your roll of pennies as a population. Use the scale of ages on the number line below and plot µ. Choose 20 pennies at random from your pile of pennies. Find the mean and standard deviation of the sample and compute a 95% confidence interval for the population mean, µ. Draw a line segment for this interval below the number line that you have scaled. Mix up the pennies and repeat the process five times. Do the intervals of your sample capture the value of µ? Why?

```
    5   10   15   20   25   30   35   40   45   50   55
___/___/___/___/___/___/___/___/___/___/___/___
```

Plot #1

Plot #2

Plot #3

Plot #4

Plot #5

12. By using the definition of outlier and the five-number summary, what is the age of a "rare" coin?

13. By using the normal curve, the mean, and the standard deviation of your sample, find the age that you would begin to save before the pennies become hard to find. Consider the coin "rare" if the age is the oldest 2% or less of the population.

As an extension, repeat this project using nickels or any other coin.

Inference for Distributions

Project 5   SNAP CRACKLE POP

Are all cereals created equal? You are to gather data on different brands and types of cereals.

1. Go to the grocery store with pen and paper in hand and gather some data on rice, oat, wheat, and corn cereals. Classify the cereal as to the first ingredient listed if it is made from a combination of grains. Record the data on a per-serving basis as follows:

   As an alternative way to gather the data, use the manufacturer's Websites for ingredients and nutritional facts.

Brand	Type	Calories	Total fat (g)	Sodium (mg)
	Rice			
	Rice			
	Rice			
	Rice			
	Rice			
	Rice			
	Rice			
	Rice			
	Rice			
	Rice			
	Oat			
	Oat			
	Oat			
	Oat			
	Oat			
	Oat			
	Oat			
	Oat			
	Oat			
	Oat			

Brand	Type	Calories	Total fat (g)	Sodium (mg)
	Wheat			
	Wheat			
	Wheat			
	Wheat			
	Wheat			
	Wheat			
	Wheat			
	Wheat			
	Wheat			
	Wheat			
	Corn			
	Corn			
	Corn			
	Corn			
	Corn			
	Corn			
	Corn			
	Corn			
	Corn			
	Corn			

Assume that the data you have collected is normally distributed for all cereals.

2. Perform a hypothesis test for the difference of the two means for each of the following questions. List the H_o, H_a, give the observed value of the test statistic, P-value, and write a word conclusion.

   a) Is there a difference in the calorie count per serving for the oat and rice cereals?

b) Do the rice cereals have fewer fat grams per serving than the oat cereals?

c) Do the rice cereals have more sodium (mg) per serving than the oat cereals?

d) Is there a difference in the calorie count per serving for the wheat and corn cereals?

e) Do the wheat cereals have fewer fat grams per serving than the corn cereals?

f) Do the wheat cereals have more sodium (mg) per serving than the corn cereals?

3. By using hypothesis testing procedures, determine if the rice cereals have a statistically significant fewer calorie count than each of the other grains. Use $\alpha = 5\%$.

Inference for Distributions

Project 6 MOVIN' OUT

Do all comparable homes cost the same in various cities? You are to gather and compare data on the same type of home in two different cities.

Mandy Middletown has just been transferred from Kansas City (from zip code 66212) to Washington D.C. (to zip code 22191). Mandy would like to find a similar home to the one in which she is presently residing, which has three bedrooms and two baths.

1. Use these zip codes to find a listing of 30 homes for sale in each of the two zip codes. Go to www.realtor.com to search for home prices.

Zip code 66212:

Zip code 22191:

2. Give the mean and standard deviation for the prices in each zip code:

   Zip code 66212:   Zip Code 22191:

   Mean = _____   Mean _____

   S.D. = _____   S.D. = _____

f) Do the wheat cereals have more sodium (mg) per serving than the corn cereals?

3. By using hypothesis testing procedures, determine if the rice cereals have a statistically significant fewer calorie count than each of the other grains. Use $\alpha = 5\%$.

Inference for Distributions

Project 6 MOVIN' OUT

Do all comparable homes cost the same in various cities? You are to gather and compare data on the same type of home in two different cities.

Mandy Middletown has just been transferred from Kansas City (from zip code 66212) to Washington D.C. (to zip code 22191). Mandy would like to find a similar home to the one in which she is presently residing, which has three bedrooms and two baths.

1. Use these zip codes to find a listing of 30 homes for sale in each of the two zip codes. Go to www.realtor.com to search for home prices.

Zip code 66212:

Zip code 22191:

2. Give the mean and standard deviation for the prices in each zip code:

    Zip code 66212:   Zip Code 22191:

    Mean = _____   Mean _____

    S.D. = _____   S.D. = _____

3. Is there a difference in the mean home prices of the two areas? Perform a hypothesis test for the difference of the two means.

   H_o:

   H_a:

   T-value:

   P-value:

   Word conclusion:

4. Approximately how many dollars in home value will Mandy loose or gain on her move?

Inference for Regression

Project 7 SHORT OR TALL

Do tall people have big feet and short people have small feet? Your task is to generate a set of data values for a person's height in inches and their shoe size.

1. In the chart below, list at least 15 people and their data. You are to standardize the sizes for male and female (female size minus 1.5 = male size).

Name	Height	Shoe size

2. Make a scatterplot of the data where shoe size (Y) depends on the height (X).

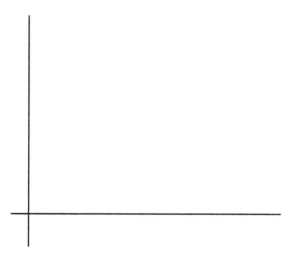

3. Find the line of best fit to the data. Using your TI-83/84, enter the height in L1 and the shoe size in L2. Push STAT, calculate, then down arrow to LinReg(*a* + *bx*). Write the equation.

4. Make a residual plot of the data here. On your calculator, choose STAT PLOT, then under type choose scatterplot. For Xlist, choose L1 and for Ylist, choose RESID. The RESID is located under LIST, NAMES.

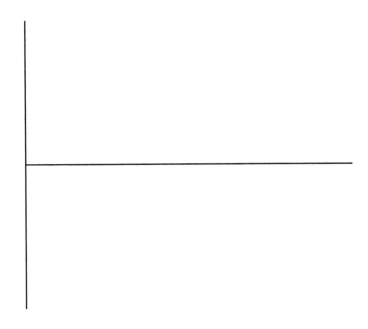

5. What does your residual plot indicate to you about the model?

6. Perform a hypothesis test to determine if there is a significant positive slope to the line of best fit. You may use the LinReg t test menu on your TI-83/84. Write the hypotheses, state the T-value, the P-value, and state your word conclusion.

   H_o:

   H_a:

   T-value:

   P-value:

   Word conclusion:

7. Suppose Sue Z. Normal walks into your classroom. Find a point estimate for the shoe size of Sue Z. Normal, who is 70 inches tall.

8. Find a 95% confidence interval for Sue Z. Normal's shoe size.

9. You may not have had anyone in your sample that was over 7 feet tall, but there are 7 footers around, particularly on pro basketball teams. Predict the shoe size for Mr. Guy Tall, who measured in at 7 feet 2 inches for Coach Him Self at the basketball tryouts.

10. Find a 95% prediction interval for Mr. Guy Tall's shoe size.

11. State why the prediction interval for Mr. Guy Tall is wider than the confidence interval for Sue Z. Normal.

Statistics for Quality

Project 8 **SKITTLES MACHINES**

You have just been hired as the quality control engineer for Skittles. Your job, because you decided to accept it, is to watch the process of the bagging of the fun-size bags to ensure that they are meeting the company's quality standards in the following ways:

A. Is the number of candies in the bag within the company's guidelines?
B. Is the color of each candy within the proportion guidelines?

Assume that your bags of fun-size Skittles are being selected randomly from the bagging machine at 1-hour intervals. Select the bags one at a time and count the data to assess if the machine is functioning properly. You are to consider the following three methods of control:
   i. Any point outside three standard deviations from the mean
   ii. Any nine consecutive values on the same side of the mean
   iii. Any two of three points more than two standard deviations on the same side from the mean

1. In the space below, make a quality control chart for the number of Skittles in the bag. Company standards are for the bags to have a normal distribution with mean 18.2 and standard deviation 0.6. Experience has shown that the machine produces bags with equal proportion of the five colors. Label the control values of the vertical axis. Plot the number of candies in each bag, in order selected, on the horizontal scale.

1 2 3 4 5 6 7 8 9 10 11 12 13 14
Bag

Use the chart above to determine if the process is in control or indicate the point where you have determined it to be out of control.

2. The bagging machine is fed by machines that produce each color of the candy. You wish to determine if each of them is remaining in control throughout the day. You are to make a control chart for the proportion of each color. Assume that the colors are of equal proportion ($p = .2$). Make the control limits for the proportion and then plot the proportions from each bag by color as you sample.

Red

1 2 3 4 5 6 7 8 9 10 11 12 13 14
Bag

Purple

1 2 3 4 5 6 7 8 9 10 11 12 13 14
Bag

Orange

1 2 3 4 5 6 7 8 9 10 11 12 13 14
Bag

Yellow

1 2 3 4 5 6 7 8 9 10 11 12 13 14
Bag

Green

```
1  2  3  4  5  6  7  8  9  10 11 12 13 14
                    Bag
```

3. Write your conclusions about the performance quality of the bagging machine and each of the color production machines in the space below as a report to your boss on the day's activity.

   A) Bagging machine

   B) Red

   C) Purple

D) Orange

E) Yellow

G) Green

Statistics for Quality

Project 9   NBA

You are to select a team from the NBA to do the following study by making a control chart for your team. As an avid fan of basketball, you have decided to chart the points scored by your team, the _____. You may use the Website www.nba.com to find the data for this project as a retrospective study, or chart the next 15 games that your team plays by checking the papers on a daily basis.

Past experience has shown that NBA teams average 98 points per game with a standard deviation of 4 points per game. You expect your team to be consistent. Consider the following three measures of them experiencing a significant change:
    i. Any game outside three standard deviations of the mean
    ii. Any nine in a row on the same side of the mean
    iii. Any two of three more than two standard deviations on the same side of the mean

Team _____

1 2 3 4 5 6 7 8 9 10 11 12 13 14 15
Game

List the date for each game for reference.

Write a summary of your team's performance over the past 15 games. Note any trends that appear on your graph. If anything happened to any player on the team during the time, or if any other significant event(s) took place that could have accounted for the change in team performance, please reference it in your report.

Looking at Data Distributions

Project 10 WHAT'S THE COST OF LUNCH?

How much does a student spend on lunch during school days? The student may choose to go to the local franchise of Betsy's Burgers, go through the a la carte line, or eat the Type A regular lunch. Any way you slice it, you still are paying for your lunch.

1. You are to gather data by using a convenient sample of 20 males and 20 females today at lunch and record the data below. Now remember this is not a scientific study, so you can take their word for what they spent for lunch and whether or not they purchased it on or off campus.

Student Count	Female On / Off	Amount $		Male On/ Off	Amount $
1					
2					
3					
4					
5					
6					
7					
8					
9					
10					
11					
12					
13					
14					
15					
16					
17					
18					
19					
20					

2. Make a stemplot of the 40 values in the space below:

3. Are there any outliers in the data set? If so, list them and give a reason to either eliminate them or keep them in the data set.

4. Find the five-number summary of the costs from the above data.

   Low = ____ Q1 = ____ Median = ____ Q3 = ____ High = ____

5. Now let's look at the data as different sets. Make a back-to-back stemplot of the data by gender in the space below.

           Female   Male

6. Make side-by-side boxplots of the Female and Male data sets in the space below.

7. How do the two distributions compare?

8. Now look at the data of the costs spent for lunch on and off campus. Make a back-to-back stemplot of the data in the space below.

                On campus   Off campus

9. Make side-by-side boxplots of the data sets above in the space below.

10. How do the two distributions compare?

11. By looking at the stemplots and boxplots you have displayed in this project, which display gives you the clearer picture for the comparisons and why?

Looking at Data Distributions

Project 11 WHAT'S IT WORTH?

What is the cost of a home in your neighborhood? Houses have different kinds of characteristics that affect price. In this project, we are going to describe a home and determine the price in your neighborhood for either the home you live in or that of a friend.

1. Describe the house you intend to research as follows:
    Number of bedrooms   _____
    Number of bathrooms   _____
    Zip code of the home   _____

2. By using the Internet and the criteria listed above, search for 30 comparable home prices for sale. An example of a Website to use is www.realtor.com. More than 30 listings will likely appear. Use a sampling technique to create a random sample of the 30 values. List the sampling technique you have chosen: _____.

3. List 30 home values:

4. Find the mean and standard deviation of the above 30 home prices.
    Mean = _____
    S.D. = _____

5. Find the five-number summary for the data above.

    Min _____ Q1 _____ Med _____ Q3 _____ Max _____

6. Which descriptive measures listed in Questions 4 and 5 are most appropriate for this distribution and why?

7. Make a boxplot of the data.

8. Go back to the Website you used and insert the address of the home you are researching. Find a minimum of five homes that have sold recently in the neighborhood.

   _____   _____   _____
   _____   _____   _____
   _____   _____   _____

9. Find the average price of the homes listed above using the descriptive measure used in Question 6.

10. What is the approximate value of your home? _____

Looking at Data Relationships

Project 12 TEST YOUR MEMORY

1. How well do you know the ages of the stars and celebrities that you hear about or see on a regular basis? You are to complete the following table with what you believe to be each person's age. After you have completed the list, you will be given their actual ages.

Name	Actual age	Your guess age	Residual	Absolute value of residual	Residual squared
George W. Bush					
Catherine Zeta Jones					
Garth Brooks					
Whoopi Goldberg					
LaBron James					
Julia Roberts					
Bill Gates					
Queen Elizabeth II					
Arnold Schwarzenegger					
Brad Pitt					
Hillary R. Clinton					
Julie Andrews					
Justin Timberlake					
Dakota Fanning					

2. Now that you have filled in your guess, your instructor will give you their actual ages.

3. To see how you have done, you will analyze the differences between their actual age and your guess. In the space below, draw a coordinate axis system and plot the ordered pairs (X, Y) where X is the actual age and Y is your guess.

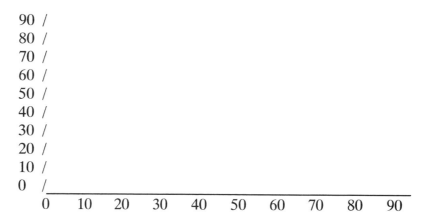

4. Now sketch the line $Y = X$. If your points fall on the line, then you have correctly guessed the ages. If your point is not on the line, then you are to draw a vertical line from your point to the line. The distance of each of these segments is a residual for that guess. Subtract the actual age from your guess to get the residual. List these directed distances in the residual column next to the guess.

5. Did you tend to underestimate, overestimate, or bounce around the line?

6. Sum the residual column. Does your answer tell you how well you guessed the ages overall? Why or why not?

7. There are two choices to consider arriving at a meaningful sum for this data.

    a) Take the absolute value of each value and then find the sum.

    _____

    b) Square each residual and then find the sum of the squares.

    _____

8. Determine which class member has the lowest:

    a) Sum of absolute value _____

    b) Sum of squares _____

9. Now use your TI-83/84 to do this problem. Use the STAT menu to enter the actual age of the star in L1 and your guess in L2.

    o In the STAT PLOT menu, make a scatterplot of the data using L1 as the Xlist and L2 as the Ylist. Choose ZoomStat to make a user-friendly window.

    o Use the Y= menu to enter the equation $Y_1 = X$ and press the GRAPH key.

    o You should now "see" the graph that you made in the above space.

    o Return to the STAT EDIT menu and define L3 as L2 – L1.

    o Define L4 as ABS(L3).

    o Define L5 as $L3^2$.

    o The LIST menu is found by pushing 2nd STAT. Now use the MATH menu to find:

        a) Sum (L4) = _____

        b) Sum (L5) = _____

10. Now that you have completed the problem above, write your definition of residual in the space below.

Producing Data

Project 13 WHAT DO STUDENTS DRIVE?

What do the students drive to school? Your job, should you decide to accept it, is to design an experiment, gather data, and answer this question.

1. Adjacent to the school, there are parking areas. You are to design a method that would produce a representative sample of 30 cars. You may use a block design, a stratified sample, or a simple random sample. State your design in the space that follows.

2. List the specific location for each car that you have selected to include in your sample before you venture into the lot (example: front lot, row 3, car #17).

3. List the results of your sample in the table that follows. You are to identify one more variable to add to the list of characteristics that are given. Write it in the space provided as "Other."

Car	Make	Year	Model	Color	Other
1					
2					
3					
4					
5					
6					
7					
8					
9					
10					
11					
12					
13					
14					
15					
16					
17					
18					
19					
20					
21					
22					
23					
24					
25					
26					
27					
28					
29					
30					

4. Summarize the data from your sample to describe a typical car in the lot.

Make _____

Year _____

Model _____

Color _____

Other _____

Producing Data

Project 14 JELLYBEAN

How many jellybeans are in the bowl?

1. Estimate the number of jellybeans in the bowl and write your guess here.

2. Now it's time to do some "fishing." Each person in the room may use a utensil (the fishing pole) provided by the teacher to remove some jellybeans from the jar. (The total number removed cannot exceed the number of black jellybeans that you have available.)

3. Replace the jellybeans that were removed with the same number of black jellybeans. (This is similar to the procedure of Wildlife Conservation Agents when they capture, tag, and release wildlife.) Write the number of replacement here.

4. Stir the jellybeans to mix in the black ones.

5. Use the fishing device to remove a sample of jellybeans from the bowl as in a simple random sample.

6. Count the jellybeans that you have captured.

    Number in total _____   Number of black _____

7. Return your sample to the jar and have each person or group repeat the same sampling method and record their counts.

8. Each person or group should write a proportion and solve for the total number of jellybeans in the jar.

9. List all the predictions made in the class and find the mean of all the values.

10. Now let's see how close we are with our predictions. Everyone participate: Divide up the total number of jellybeans and count them! Write down the total number of jellybeans here. How close were the predictions to the actual count?

11. The Missouri Conservation Commission is to determine the size of the Canadian Goose population in the state of Missouri. Design a capture-recapture procedure they could use to estimate the size of the flock.

Probability: The Study of Randomness

Project 15 FAIR COIN

Let's flip a coin and record the outcome. As a group, discuss the following items before you begin the experiment:
- How to flip the coin so it is done in the same manner each time.

- Whether or not to turn the coin when you catch it.

- Which side of the coin should start facing up? Alternate heads or tails up, or always the same?

- What do you do if you drop the coin? Will you count the toss or re-flip?

1. You are to flip your chosen coin according to the agreed-upon guidelines above. After each flip, you are to record the results as an H (head) or T (tail). Keep track of the total number of heads after each flip and find the percentage of heads after each flip.
Record the results on the grid that follows:

Trial number	Start up H or T	Outcome H or T	Total number heads	% heads
1				
2				
3				
4				
5				
6				
7				
8				
9				
10				
11				
12				
13				
14				
15				
16				
17				
18				
19				
20				
21				
22				
23				
24				
25				

2. The preceding data give you a numerical representation of the behavior of your coin. Make an observation about the percent heads column.

3. Let X represent the trial number and Y the percent of heads after X trials. Plot the set of (X, Y) on the grid below.

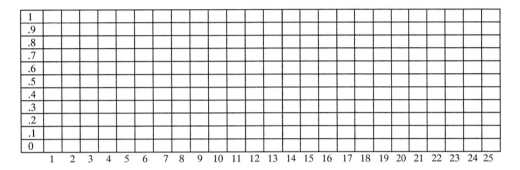

4. Make a line plot of the points on the above grid by connecting the points as you move from left to right.

5. The graphical display of the data on the grid gives you a picture of the behavior of your coin over time. What does your graph indicate to you about your coin?

6. By using both the numeric display and the graph, make a conjecture about your coin.

7. If the other members of the group used different coins, compare the graphs of the various coins. State your observations here.

   Penny

   Nickel

   Dime

   Quarter

8. Using what you have learned, consider the following. To start a volleyball match, the official begins with a coin flip. Do you think that you can call it correctly more than 50% of the time if you see the top side of the coin and know that she will not turn it over? Review your results of the coin flip in column 2 and column 3. State the proportions below.

If start heads, then end heads	_____
If start heads, then end tails	_____
If start tails, then end heads	_____
If start tails, then end tails	_____

9. Is this a fair method of starting the match?

From Probability to Inference

Project 16 ODD OR EVEN

Part 1:
The following grid consists of five rows and nine columns. You are to place a penny on the square marked START. With each throw of the die, you will move down one row and left one square if the roll is odd, or down one row and right one square if the roll is even. One game consists of four rolls of the die. Each game will place your penny at some position on the bottom row, and when that happens, place a tally mark on that square. You are to play 16 games.

But wait! Before you begin playing, you are to predict the final outcome of the 16 games by placing numbers on the lines below the grid that you anticipate as your final result. The sum of your predictions should equal 16.

Now you may play the game. Let the good times roll!

				START				

___ ___ ___ ___ ___ ___ ___ ___ ___

1. So how did you do with your predictions? Were you surprised at your results? Discuss symmetry and the cell locations that are impossible.

2. The model used in this game is called the binomial distribution. List four characteristics of a binomial model in the space below.
    a.

    b.

    c.

    d.

3. Use the binomial model to find the probability of ending on each of the five possible final positions of the game board on the previous page.

    You may use the binomialpdf function on the TI-83/84 or the binomial theorem to do your work. Number the final positions 1, 2, 3, 4, 5 from left to right on the bottom row and give the probability of ending in that position.

    Position 1 _____

    Position 2 _____

    Position 3 _____

    Position 4 _____

    Position 5 _____

4. Compare the results of the game to the result using the probability as computed above. List your observations in the space provided.

Part 2:

Repeat the game with the following changes in the movement rule. Roll a die and move down one row and left one space if the roll is 1 or 2, or move down one row and right one space if the roll is 3, 4, 5, or 6. List your results here.

Position 1 _____

Position 2 _____

Position 3 _____

Position 4 _____

Position 5 _____

1. How are the two games different?

2. How are the two games similar?

Introduction to Inference

Project 17 AHOY MATES

The company claims, "There are over 1000 chips in every bag." How can they make such a claim? As a consumer, your job is to design an experiment to check the claim made by the company.

1. Describe a sampling procedure you would use to gather the data. Include how to select the bags, cookies in the bag, and chips within the cookie.

2. Consider the sample size you think is needed for the problem. Choose a sample size for the number of cookies that will allow for an approximately normal distribution of the sample mean number of chips within a cookie.

3. You are now ready to begin your count. Most students consider this to be a "crummy" problem. State how you decided to "count" the partial chips?

4. Make a histogram and stemplot of the distribution of chips within the cookies.

5. Write a 95% confidence interval for the number of chips per cookie.

6. To consider the number of chips in a bag, you must now consider the number of cookies in the bag. While we have not chosen a large number of bags of these cookies, we can make an observation about the number of cookies in the bag. What is it?

7. Using the distribution of chips per cookie and the number of cookies in the bag as your guide, write a 95% confidence interval for the number of chips in the bag by making a linear transformation of the data. $\mu_{\text{chips in bag}} = n\, \mu_{\text{chips in cookie}}$ and $\sigma_{\text{chips in bag}} = n\, \sigma_{\text{chips in cookie}}$

8. What do you think about the company's claim of 1000 chips in each bag?

9. The company has just hired YOU as chief statistician. Address the problem with a hypothesis-testing procedure. State the hypotheses (H_o and H_a) for the claim made by the company. Report the value of the test statistic and the *P*-value of the test.

   H_o:

   H_a:

   T = _____   *P*-value = _____

   Word conclusion:

10. Write a summary (approximately 50 words) to the problem that you would submit to the company president.

11. Identify five items relating to statistics that you have learned from this project.

   1)

   2)

   3)

   4)

   5)

You may now eat the data. Got milk?

Inference for Proportions

Project 18   TASTE THE DIFFERENCE

Can you really taste the difference in the brown M&M's?

1. You are to select a partner to perform the following experiment. One person will record, the other will taste. After completion, you will reverse roles and repeat the experiment.

2. Take a small number (approximately five) of brown M&M's and train your taste buds by eating them slowly one at a time.

3. Now take a non-brown M&M and eat it slowly. Sorry, no more than five tries of knowing each color, brown vs. non-brown. After all, if you eat all the data, we can't run the taste test!

4. You must trust your partner on this one to record the data, as you are the lucky one that gets to eat your way through. Close your eyes; blindfolding is necessary to prevent looking at the candies.

5. The person recording should choose ten M&M's of random colors. Make sure that there is a good color mixture inclusive of three to five brown candies. One by one, you the taster should eat the candies given to you by your partner and tell him/her if you think it is brown or not brown. The partner is to record the results as correct or incorrect for each trial. Do not tell the taster if they are correct after each trial. Record the results for each of the ten trials as T = true for a correct guess and F = false for an incorrect guess in the chart below.

Trial #	True or False
1	
2	
3	
4	
5	
6	
7	
8	
9	
10	

6. This experiment consists of independent trials, each of which has two possible outcomes, success or failure, where the probability of success and failure add up to one. If a person has no idea as to the color, then the probability of success is .5 for every trial. Count the number of successes that were observed in the ten previous trials. Find the probability of that number of successes if indeed you were guessing. You may use the binomial theorem or the binomialpdf function on the TI-83/84 calculator. Express the method of your computation in the space below.

7. Make a probability density function for the random variable of the number of successful guesses in the ten trials. Use $p = .5$ and list the results in the table below.

Number	Probability
0	
1	
2	
3	
4	
5	
6	
7	
8	
9	
10	

8. Find the expected value of the number of successes for the distribution from the list above.

9. Find the standard deviation for the distribution above.

10. Let *n* equal the number of successes that you observed in your taste test. By using the standard normal distribution, find the $P(X \geq n)$. Use the continuity correction factor.

11. On the TI 83/84, choose the binomialcdf function and find the $P(X \geq n)$.

12. Compare your answers to Questions 10 and 11. Are they the same or different? Why?

13. So how did you do? Do you think now that you really can tell the difference? Are you willing to bet your leftover M&M's?

14. Switch sides of the table and repeat the taste test with your partner.

15. So who did the better tasting? _____ Repeat the experiment so that each of you have 20 trials. Let *k* equal the total successes in the 20 trials.

    Total number of successes in 20 trials = _____

16. Test the hypothesis that you are guessing verses the alternative that you can tell the difference. Use the *P*-value techniques to test your hypothesis and write a word conclusion that could be used as evidence in court to support your findings.

    H_o:

    H_a:

    $Z =$ _____

    P-value = _____

    Word conclusion:

    Now eat the evidence so that the next class is unaware of "how sweet it is" in here!

Inference for Two-Way Tables

Project 19   PLAIN AND PEANUT

What is your favorite color of M&M? Can you really taste the difference? When you buy the bag, do you eat them first or last? The real question here is to address the proportions of colors of M&M candies. Over time, the company has changed the colors and proportions. What are the proportions now? You will discover the answers to these and other questions as you proceed with this project.

1. Identify the colors of plain M&M's that are in the package. List the colors.

2. Before you begin the data collection, state the percentages of colors that you *think* are in the bag. Write your predictions next to each color on the list above.

3. Make a frequency distribution of the counts and give the relative frequency for each.

4. The "color" of M&M is what type of data? _____

5. Choose an appropriate type of graph to chart the data and state why you chose it.

   _____ graph because,

6. Graph the data by using your graph choice.

7. Repeat the procedure using peanut M&M's. What are the colors used by the company?

8. Make frequency distributions of the counts and find the relative frequencies.

9. Graph the data using the same type of graph used in the plain M&M data.

10. Compare and contrast the two distributions and graphs.

11. Go to www.MARS.com and find the color proportions per package the company intends for both the plain and peanut M&M's.

   a) Plain
      List the colors and the company's proportion claim.

   b) Peanut
      List the colors and the company's proportion claim.

12. Identify a procedure that you could use to check the goodness of fits for your data on color proportions.

13. Write the null and alternative hypothesis you would use to verify the company's claim.

    H_o:

    H_a:

14. Display in a matrix the observed and expected cell values and perform a Chi-Square analysis for the (a) plain and (b) peanut data.

    a) Plain

Color	Observed	Expected	$(O-E)^2 / E$

b) Peanut

Color	Observed	Expected	(O-E)² / E

15. State the value of Chi-Square and write a word conclusion to the hypothesis test you have performed using the 5% level of significance.

   a) Plain

   b) Peanut

16. Now consider the type of M&M and the color. Use the two samples that you counted to test the hypothesis that the type of M&M is independent of the colors. Use the following grid cells to display your data using the columns as types and the rows as colors.

Color	Plain	Peanut	Total
Total			

State the hypotheses, the value of the test statistic, the *P*-value, and write a word conclusion.

H_o:

H_a:

Chi-Square = _____

P-value = _____

Word Conclusion:

Multiple Regression

Project 20 MANY VARIABLES DO PREDICT

Can you predict your arrival time at school? While there are many variables that could impact this time, we will consider only the two variables of the distance that you are from school and the time that you leave home to go directly to school.

1. Gather information using the time in minutes starting at $t = 0$ as 7:00 AM and give the distance in tenths of a mile. Each member of the class is to record the data and return it tomorrow for class.

Record the information in the table. Use a minimum of 25 sets of values.

Arrival time	Distance	Start time

2. Let Y = the arrival time, X_1 = the distance, and X_2 = the start time to find a multiple regression equation for the relationship. You are to use a spreadsheet to enter the data and generate a multiple-regression output for the variables defined above. Excel or Minitab will output this information. Attach a copy of the output to this report.

3. Write the estimated equation from the output.

4. State the hypothesis tested by the ANOVA F statistic for this problem in words.

5. State your conclusion to the hypothesis based on the computer output.

6. What percent of the variation in arrival time is explained by the distance and start time?

7. If you live 5.3 miles from school and leave at 7:15 AM, what is your expected arrival time?

One-Way Analysis of Variance

Project 21 GROWTH GROUPS

As you wander the hallways of the school, you note various heights of students. For this project, you are to select a random sample by means of a convenient method to compare the four grade levels of students and their heights.

1. List the heights of a minimum of 15 students of the same gender from each of the four grade levels in the space provided.

Freshman	Sophomore	Junior	Senior

2. Make side-by-side boxplots for the data sets listed above.

3. Summarize the relationship of the boxplots. How are they alike and how are they different?

4. Make a normal quantile plot for the data of each of the four grades, and state your conclusion regarding the normality of each.

   a) Freshman

   b) Sophomore

   c) Junior

   d) Senior

5. Summarize the data for each class in the table below.

	Freshman	Sophomore	Junior	Senior
Mean				
Standard deviation				
Sample size				
Standard error				

6. Assume that the groups have equal standard deviation and perform the one-way ANOVA F test. State the hypotheses.

7. State the degrees of freedom for the ANOVA F test.

8. Summarize the results in the ANOVA table below.

Source	Degrees of freedom	Sum of squares	Mean sum of squares	F
Groups				
Error				
Total				

9. You may use a computer output to do the computation for this problem and attach a copy of its output to this project.

10. State the critical value of the test statistic. Use an α level of 5%.

11. Summarize the conclusion to the hypothesis test. Include the *P*-value and state your results in terms of the variables in the problem.

Analysis of Variance

Project 22 RANDOM ASSIGNMENT

In a large school where the students are assigned to their class schedule by computer, it would sound reasonable that the students within all classes of the same topic, for example, Algebra II would have a similar distribution of last names.

Your instructor has noted that in each of his classes this year the students seem to be clustered in an alphabetical grouping. Morning classes tend to have students with last names like Adams, Beltron, and Jones, and afternoon classes seem to have Smith, Thompson, and Young.

Is this true in your classes? Choose an instructor and make a distribution of the first letter of the last name for each member of the class. You are to use three classes for this project. Recall that we are checking for the assignment of students within the subject matter, so more than one section of a class must exist. Assign a numeric value to each letter of the alphabet as follows: a = 1, b = 2, c = 3, etc.

While the distribution of values will not be normal, the central limit theorem indicates that the sample means will tend to approach a normal distribution as the sample size approaches 30. Use this fact to decide if the students in the classes could be considered a random assignment to the daily class schedule.

1. Make a stemplot of the data for each class in the space below.

2. What relationships, similar or different, do you see in the graphs?

3. Make side-by-side boxplots for the preceding data in the space below.

4. Compare the plots and state you conclusions.

5. Perform a hypothesis test by comparing the means of the classes using a one-way ANOVA F test on the three distributions. Write the H_o, H_a, give the value of the F statistic, and state the P-value for this test.

   H_o:

   H_a:

   F-value:

   P-value:

   Word conclusion:

6. Write a summary that you would present to the principal of your school regarding the random assignment of students to the various periods of the daily schedule.

Research Project

Project Final

Within this project, you must demonstrate five different aspects of a statistical study. You may use either an observational study or a designed experiment. Your instructor must approve your topic of research before you begin.

1. Pose a question that you can answer with data.
   Examples:
   - Is there a difference in the normal mean body temperature between men and women?
   - Is there a higher percentage of births around the full moon?
   - Does the color of light bulb affect plant growth?
   - Does the American League hit more runs than the National League on average?
   - Do M&M's and Skittles have the same proportion of green candies per bag?
   - Is there a difference in the average word length between the writings of Henry David Thoreau and Robert Frost?

2. Explain how you will select the data or design the experiment to gather the data. Give a rationale for your choices.
   Examples:
   - Simple random sample
   - Block design
   - Systematic sampling

3. List the data values you obtain and choose an appropriate graphical display for the data. The descriptive methods are to relate to the question that you have asked. Descriptive methods include stemplots, boxplots, histograms, and pie graphs, along with a numerical summary of mean and standard deviation (normally distributed) or the five- number summary (not normally distributed).

4. Use a hypothesis-testing procedure to analyze the data. Options include the Z-Test, T-Test, and Chi-Square Test. Give supporting reasons for the choice of the procedure used. State the null and alternative hypothesis, the value of the test statistic, and the P-value of the test. Summarize your conclusion in terms of the question that you posed.

5. The final paragraph is to summarize what you have concluded and demonstrated in your work using language that a non-statistician would understand.

Statistics in Research

John C Turner
U.S. Naval Academy
Annapolis, MD

Preface for Statistics in Research

One of the current ideas in statistical education is that courses should use "real world" data wherever possible. Many texts do this by citing data from various studies. This is fine, as far as it goes, but this approach has several shortcomings. For one thing, the student gets an extremely brief view of the scientific problem. Sometimes, the only information about the problem comes from the title of the journal article. Further, the textbook presents a highly "sanitized" view of the problem. The author has read the article and extracted precisely the information that is needed to do the problem discussed in the text. At the U.S. Naval Academy, I have put together a course for students who have completed introductory statistics. In this course, the students read actual journal articles and discuss how statistics was used in the conclusions of the article. This led to a number of observations. Sometimes, necessary data (such as sample size) is buried elsewhere in the article or completely missing. Some articles used inappropriate methods or misinterpreted the results of the analysis. This includes, for example, confusing small P-values with large effects. Lastly, by considering the entire article, the students got a much better idea of how the statistical analysis related to the scientific conclusions of the article.

Since starting the course described above, I have also used articles in basic statistics, notably at The University of Texas at Austin. The students were quite good at following the statistical analysis and some could even follow analyses that had not been covered in the course. Many of the students commented that they were impressed that they could make immediate use of what they had learned, even though it was a very introductory course.

I have tried to include articles from a number of fields and I hope that the students will find several articles of interest. I have tried to use articles that do not require much specialized scientific knowledge to follow. Above all, it is my hope that by following the entire article, the student will come to appreciate that statistics is not just a bunch of formulas that you guess at on a test, but rather a method of analyzing data that is critical to much scientific research.

Introduction

One of the most interesting aspects of statistics is that it has applications in a wide variety of problems in a wide range of fields. Any situation where we wish to use sampled data to draw broader inferences is fodder for statistics. In many situations, of course, considerable background knowledge is required in order to make good sense of the problem being solved and the significance of the solution. However, there are many problems that can be addressed without much in-depth knowledge of the particular field. I have collected several such papers for use as a companion in a beginning statistics course.

The term "beginning statistics course" encompasses a wide range of courses. There is not a single, well-defined set of topics that would be included in such a course. However, there *is* a set of topics that surely would be covered by any such course and also a slightly larger set of topics, only some of which would be included in any particular course. There is also not a set sequence of these topics. The placement of such topics as contingency tables and ANOVA is somewhat variable, if they are included at all.

This leads to several issues in the use of this material. There are two broad approaches available. One is to utilize all (or many) of the articles as progress is made through the course. When covering the different types of variables, for example, students could consider all of the articles and identify the types of variables in each case. Then, when experimental design is covered, students could return to that set of articles and discuss the design in each one. This could continue throughout the course. Of course, when the course covers a specific statistical method, only the papers that use that method would be considered.

Alternatively, use of the articles could be postponed until the first statistical tests have been covered. Then, students might consider some of the articles that use the methods studied thus far. For these articles, students could then identify all the variables in these articles, as well as discuss the sampling scheme and experimental design, etc. With this approach, when a new method is covered, students would repeat the exercise with an article that makes use of the new method.

A third approach would be a combination of the two above. Early in the course, before statistical methods have been covered, students would use only some of the articles and discuss variables, sampling, and design. Then, when a particular method is covered, students can perform the full analysis of a paper that uses the method just covered. If this article is one that has previously been discussed, then students will have already covered the notion of variables and design. If the appropriate paper for this method has not yet been covered, then students will begin with the identification of variables, etc.

Reaching over all these concerns is the interests of the students. I have tried to include a variety of fields, for several reasons. A given class will find some articles more familiar or more interesting than others. The instructor will do well to bear this in mind. At the same time, it is possible to make use of articles outside those presented here. The instructor should take particular care in selecting additional articles to see that the articles are appropriate both in the scientific problem being addressed and in the statistical approach. There is also a decision in terms of whether to use articles that illustrate poor uses of statistics. There are many such examples, but I have chosen not to include any in this supplement. Nonetheless, students may learn quite a bit from a poor application of statistics if the example is carefully chosen and it is made clear that the point is for students to learn what to avoid rather than to simply criticize of the article.

Common Tasks for All Articles

1. Read the article. Some of the terms may be unfamiliar, but you will have to decide which terms are crucial and which are not. For instance, it is enough to know that an article deals with a specific species of fish. You don't need to know that many details about that species. It is probably a good idea to read the article all the way through first. Then review the general flow of the article. Some articles start with one problem, but then discover something new partway through. Try to see all the parts and see how they fit together. In many experimental papers, the first section will demonstrate that there are no significant differences between the treatment and the control groups on important variables. Thus, if differences are noted after the treatment, it is logical to consider them as resulting from the treatment rather than from some initial differences between the groups.

2. Determine the variables in the study. There may be quite a few, especially in cases where the article has multiple objectives. Which variables are quantitative and which are categorical? Some texts discuss a third type of variable, ordinal. These are similar to categorical in that you can't do arithmetic on them, but they have the additional property of a natural ordering. A common example is where responses to a survey are "Strongly disagree," "Disagree somewhat," etc. Data like this are usually converted to 1–5 or 1–7 (called a Likert scale) and then considered to be quantitative. It is less clear what to do with categories such as Young, Medium, and Old.

3. Very few journal articles have graphical displays of the data. You should consider what is lost by not having these. Graphical displays are very helpful in giving an overview of the data to start with. In some types of analysis, the assumption is that the data have a normal distribution, at least approximately. A histogram may suggest how good this assumption is. The student should also bear in mind that a journal article does not show everything. Except for a few rare cases, you will not see all the data. You can at least hope that someone looked at the appropriate graphics before doing the analysis. Are there certain variables that you suspect are not normally distributed? How important is this to the results?

4. Sampling and experimental design are issues that are sometimes discussed explicitly, sometimes implicitly, and sometimes not at all. If the article reports on an observational study, what "lurking variables" might be present? Did the authors make any note of them and try to correct for them? (You might not know how to correct for these other variables, but there are methods that can be used.) In a matched-pairs experimental design, why was the matching necessary? How would the authors go about doing the

matching? Also, was the experiment "double blind"? It is not possible to make some experiments blind. How might this affect the results?

5. The instructor and the student alike need to be aware that there is more than one way to achieve the same result. Many texts (correctly) discuss the Z-statistic for testing two proportions. This same (two-sided) test can be done with a χ^2 statistic with 1 df. Similarly, the two-sample t test can also be done using an F statistic with 1 df in the numerator if you consider the two-sample t test as a simple ANOVA. The (two-sided) P-values are exactly the same for each of the equivalent methods. For the student, the important thing is to consider what test *you* think should have been done. If enough data is given, you can reproduce the test and compare your P-value to the one given.

6. As you read through the analyses, consider the number of sides and the P-value. It is quite common to find authors doing two-sided tests when a one-sided test is more appropriate. What is the effect of using the wrong number of sides? Is this effect the same when the P-value is quite small as when it is more modest? It is common to find that articles only state that the P-value is less than 0.001, say, and not give the exact value. Why do they not give the exact value? Does this affect the usefulness of the article from the point of view of the intended audience?

7. You will also notice that some articles only test hypotheses, some only give confidence intervals and some do both. What are the advantages and disadvantages of each method (tests and intervals)? In a given setting, does one approach give a better (more useful) answer than the other? Are there problems where it is not feasible to give a confidence interval?

8. Did the article stop short of truly answering their question? In ANOVA and contingency tables, we can test if all groups are equal. If they are not all equal, it would seem reasonable to consider which groups are different from which. This is not a simple matter but would be a good approach to the question of interest.

9. It is important to repeat the calculations in the article whenever possible. This makes it clearer what the authors actually did. This can show you that scientists are obtaining actual results using methods that you covered in your course. It can also point out typos (which are more common than you might think). Sometimes, it can also clarify what calculation the authors actually did. The article might not be clear on whether the P-value is one- or two-sided. By performing the calculations, you can determine for yourself which one it is. Furthermore, doing the calculations yourself helps you see what values are needed. In more than a few articles, the value of the sample size(s) can be hidden many pages away from the calculation or missing entirely. It is not uncommon to find that, while averages are given, the standard deviations are not, making it impossible to reproduce the calculation. On this point, note that some articles will give standard deviations and some will give standard errors, the standard deviation divided by the square root of the sample size. Sometimes, the terminology in the article is unclear and sometimes it is simply wrong. You will find out once you try to do the calculations yourself.

Learning What Not to Do

There are regrettably many examples of poor statistical practice to be found in journal articles. One of the purposes of publishing scientific articles is to spread information. There is not enough space in a journal article to cover every detail of the research. At the same time, it is important to give some support for the conclusions of the article. One part (and only one part) of this support comes from statistics. Students should consider the statistical evidence in a paper in this light: Does it convince the reader that the conclusions of the paper are right?

Students often have trouble seeing this perspective when they read an article. However, one semester I had a student who was a female soccer player. She reported on an article that suggested a change in training methods to reduce ACL injuries (which are an increasing problem among female athletes). After she gave her report, I asked her, "Based on this article, would you change your training methods?" She thought for a moment and then gave a sincere explanation of why the evidence in the article was not convincing to her.

There are a variety of common omissions in journal articles and some are more serious than others. I have listed some of the possible omissions and a few ideas on how important each may be.

1. Failure to specify N. When N is omitted, the reader is completely at the mercy of the authors to interpret the statistics given. I would be very skeptical of a paper that would not include such basic information.

2. Failure to give standard deviations. When dealing with means, omitting standard deviations is the same as not giving the units of the data. I would again be skeptical.

3. Failing to distinguish between standard deviations and standard errors. Some authors appear not to know the difference between the two. Others will simply include one or the other and not be clear about which one is given. Sometimes, you can "back calculate" and determine which value was given. For instance, if a difference is claimed to be significant, you can calculate the t statistic, assuming on the one hand that the given value is a standard deviation and on the other hand assuming the value is a standard error. If one is significant and the other is not, then you know whether the value is a standard deviation or a standard error. If they are both significant, then you are still lost.

4. Failure to graph the data. Research journals have very few graphs that look like ones in statistics texts. The text should make it clear that an important purpose of these graphs is in the initial phase of the analysis. We hope that the authors of the journal article did similar graphs when they began their analysis.

5. Do the data appear to be normal? This is similar to the previous point. We hope the authors did some consideration of this problem. Also, students should remember that many of the methods (t tests in particular) are rather insensitive to deviations from normality. We generally need only that the distribution be mound-shaped.

6. ANOVA and t tests when the standard deviations are quite dissimilar. Most texts on ANOVA stress the need for at least similar standard deviations (within a factor of 2, typically). Many texts do not make a similar claim for the t test, even if the text points out how the t test is a special case of ANOVA. An article I chose not to include had done ANOVA where the largest standard deviation was over 1000 times larger than the smallest standard deviation!

7. Extremely large df in t tests. Most texts teach that the t distribution is very close to normal for df > 30. However, there are journal articles that report t values with df in the thousands or tens of thousands. This is not incorrect, but is still disturbing. To me, it sounds like the author considers that statistics is a necessary evil for getting an article published, rather than a source of information about the data.

8. Simple typos. These are not serious and I have not seen any cases where typos have altered the conclusions of the paper. Further, any critic of typos is doomed to make one himself. They can be interesting for the student to catch and demonstrate that they know what some of the values should be. One example I have seen includes a list of t statistics where two of the values were identical but had different significance levels (for the same N). Enough other data was given to determine which value was right and which was wrong. The significance level given was correct and the typo turned out to be in the statistic itself.

The Articles

1. "Sarcopenia, calf circumference, and physical function of elderly women: A cross-sectional study," Yves Rolland, et al, *Journal of the American Geriatrics Society*, 51:1120–1124, 2003.
2. "Opinion formation in evaluating sanity at the time of the offense: An examination of 5175 pre-trial evaluations," Janet I. Warren, et al, *Behavioral Sciences and the Law*, 22:171–186, 2004.
3. "Imagination, personality, and imaginary companions," Tracy R. Gleason, Raceel N. Jarudi, and Jonathan M. Cheek, *Social Behavior and Personality*, 31(7):721–738, 2003.
4. "Influence of seeding depth and seedbed preparation on establishment, growth and yield of fibre flax (Linum usitatissimum L.) in Eastern Canada," S. J. Couture, et al, *Journal of Agronomy & Crop Science* 190:184–190, 2004.
5. "Shape of glass and amount of alcohol poured: Comparative study of effect of practice and concentration," Brian Wansink and Koert van Ittersum, *BMJ*, 331:1512–1514, 2005.
6. "Linking superiority bias in the interpersonal and intergroup domains," Matthew J. Hornsey, *The Journal of Social Psychology*, 143(4):479–491, 2003.
7. "Monitoring early reading development in first grade: Word identification fluency versus nonsense word fluency," Lynn S. Fuchs, Douglas Fuchs, Donald L. Compton, *Exceptional Children*, 71(1):7–21, 2004.
8. "Low-fat dietary pattern and risk of invasive breast cancer: The women's health initiative randomized controlled dietary modification trial," *JAMA*, 295(6):629–642 (online), 2006.

9. "Association of long-distance corridor walk performance with mmortality, cardiovascular disease, mobility limitation and disability," Anne B Newman, et al, *JAMA,* 295(17):2018–2026 (online), 2006.

"Sarcopenia, calf circumference, and physical function of elderly women: A cross-sectional study," Rolland, et al, *Journal of the American Geriatrics Society* 51:1120–1124, 2003.[1]

As is stated on the first page of the article, sarcopenia is "the involuntary loss of skeletal muscle mass that occurs with advancing age." One of the concerns about sarcopenia is that it leads to a loss of autonomy in the elderly, meaning that it makes it difficult for the elderly to get about and care for themselves.

The current method for measuring muscle mass is based on X-rays (DEXA). This is not simple and is expensive. The authors seek a simpler method for detecting sarcopenia. They decided to see if calf circumference (CC) could be used to indicate sarcopenia.

1. On p. 1122 of the article is a table giving statistics on various measurements for the subjects in the study. What distribution is implied for these measurements? Does this seem a reasonable choice for the distribution?
2. The same table gives the Minimum and Maximum for the measurements. Are these columns useful? How does the sample size affect the usefulness?
3. How were women recruited for the study? Does this pose any issues for their conclusions?
4. At the end of p. 1121, the authors give the mean (average) and SD for age. 84% of the patients were between 75 and 85. Does this seem about right?
5. On p. 1122, the article states that 9.5% of the subjects had sarcopenia. Confirm their confidence interval. What assumptions are they making in calculating this confidence interval?
6. Table 3 on p. 1123 gives risks for various outcomes with sarcopenia defined using SMM and using CC. Earlier in the article, the authors discussed using SMM<5.45 as a definition of sarcopenia. What is the basic idea for this cutoff?
7. Table 3 gives odds-ratios (OR) for the various outcomes. These are "adjusted" using a more advanced technique called logistic regression. Nonetheless, the interpretation for adjusted OR is the same as for non-adjusted OR. The odds of some event is defined as $p/(1-p)$. What two odds are used in the ratio of odds? Which odds is in the numerator? Why does it matter?
8. If sarcopenia was not a factor for one of the outcomes, what would you expect OR to be? How would you interpret larger values of OR? Smaller values?
9. Consider the *P*-values for sarcopenia defined using SMM. What would you say overall about these? How does this relate to the column of confidence intervals? Since one of the concerns about sarcopenia was how it would affect the lives of the elderly, what would you conclude about using SMM to define sarcopenia?
10. Now consider the part of Table 3 that refers to using CC to define sarcopenia. What do you make of the *P*-values? How do these relate to the confidence intervals? How does this relate to using the CC definition of sarcopenia?
11. Table 2 on p. 1122 gives correlations between ASM and other measurements. What do you make of the negative correlation with Age? What about all the other

---

[1] © 2003, Blackwell Publishing. Used by permission.

positive correlations? Are we sure that all these correlations are not simply the result of chance?

12. In the last full paragraph in the first column of p. 1122, it says that there is no significant difference in ASM between different groups. What method was (probably) used to determine this?
13. In the top two paragraphs in the second column of p. 1122, the authors discuss a multiple regression model for ASM. Should we be surprised by the variables that arise in this model? One of the coefficients is surprising—which one? In light of Table 2, how can this occur?
14. It is said that "48% of the variance" can be explained by the model. Where would you find this on the regression output?

On p. 1122 is a section titled "Best Discriminatory CC for Determining Patients with or without Sarcopenia." This uses terminology that might not be familiar. Suppose we have a test for some medical condition. The test can indicate Positive or Negative. Unfortunately, most tests are not 100% accurate. Some people who have the condition will test Negative and vice versa. A False Positive (FP) occurs when someone without the disease receives a Positive test result. A True Positive (TP) occurs when someone who does have the disease (correctly) receives a Positive test result. Many tests are based on a cutoff value for some quantity. Different values of the cutoff will produce different probabilities for FP and TP. The receiver operating characteristics (ROC) curve considers different cutoffs and graphs the probability of a FP as x against the probability of a TP as y.

For a given value of the cutoff, we can define sensitivity and specificity. Sensitivity is the probability that a person with the condition will receive a Positive test result. Specificity is the probability that someone without the condition will receive a Negative test result. Note that we would like both of these values to be large. Along with these is PPV and NPV. PPV is Positive Predictive Value, the probability that someone with a Positive test actually has the condition. NPV is Negative Predictive Value, the probability that someone with a Negative test actually is free from the condition. The relation between the test characteristics (sensitivity and specificity) and PPV and NPV depends on the incidence rate of the condition.

15. In this study, what is the incidence rate of sarcopenia?
16. Use the value of incidence rate to confirm the values for PPV and NPV.
17. The paper says that the ROC gave an optimal value of the cutoff for CC as 32 cm, but the authors decided to use a cutoff of 31 cm. If CC is above 31 cm, what would this "test" say about sarcopenia?
18. The sensitivity of the test (using 31 cm) is 44.3%. Sketch a graph of CC and show this probability. Should you graph CC for those with sarcopenia or those without sarcopenia?
19. The specificity is 91.4%. Show this on a graph.

20. There is quite a disparity between sensitivity and specificity for this test. Which of these is more important depends on how the test will be used. How will this test be used and, consequently, which one should be larger?
21. We do not know the SD for CC for those with and without sarcopenia, but the overall SD is shown as 3.1 in Table 1. If we assume the SD for both groups is 3.1, find the mean CC for each group, based on the sensitivity and specificity.
22. Generate the ROC for cutoffs from 28 to 35 cm.
23. The article says that they considered ROC curves including other variables. What would the cutoff for such a test look like?

Sarcopenia, Calf Circumference, and Physical Function of Elderly Women: A Cross-Sectional Study

Yves Rolland, MD,*† Valérie Lauwers-Cances, MD,‡ Maxime Cournot, MD,‡ Fati Nourhashémi, MD,* William Reynish, MD,* Daniel Rivière, MD,† Bruno Vellas, MD,* and Hélène Grandjean, MD‡

OBJECTIVES: To determine whether calf circumference (CC), related to appendicular skeletal muscle mass, can be used as a measure of sarcopenia and is related to physical function.

DESIGN: Retrospective analysis of data from 1992 to 1994 of the European Patient Information and Documentation Systems Study.

SETTING: Community setting in France.

PARTICIPANTS: One thousand four hundred fifty-eight French women aged 70 and older without previous history of hip fracture were recruited from the electoral lists.

MEASUREMENTS: Muscular mass was assessed using dual-energy x-ray absorptiometry (DEXA). CC was measured using a tape measure. Anthropometric measurements (height; weight; and waist, hip, and calf circumference), strength markers (grip strength), and self-reported physical function were also determined. Sarcopenia was defined (using DEXA) as appendicular skeletal muscle mass (weight (kg)/height (m²)) less than two standard deviations below the mean of a young female reference group.

RESULTS: The prevalence of sarcopenia was 9.5%. CC was correlated with appendicular skeletal muscle mass ($r = 0.63$). CC under 31 cm was the best clinical indicator of sarcopenia (sensitivity = 44.3%, specificity = 91.4%). CC under 31 cm was associated with disability and self-reported physical function but not sarcopenia (defined using DEXA), independent of age, comorbidity, obesity, income, health behavior, and visual impairment.

CONCLUSION: CC cannot be used to predict sarcopenia defined using DEXA but provides valuable information on muscle-related disability and physical function. J Am Geriatr Soc 51:1120–1124, 2003.

Key words: disability; physical function; calf circumference; elderly; sarcopenia

The involuntary loss of skeletal muscle mass that occurs with advancing age is called sarcopenia.[1] Sarcopenia also results in a decrease in muscular strength and endurance and is significantly associated with loss of autonomy in the elderly.[2,3]

Few studies have provided a description of the anthropometry and body composition of subjects aged 60 and older.[2,4-6] There is also little information relating to the prevalence of sarcopenia and its implications, mainly because of practical difficulties in the study of the muscle. Dual-energy x-ray absorptiometry (DEXA) is a valid method[4,7] of quantitative assessment of the skeletal muscle mass in vivo. The muscular mass of the four limbs, or appendicular skeletal muscle mass (ASM), can then be obtained. This technique to assess muscle mass is currently only used as a research tool.

In 1996, the National Institute on Aging concluded that it was necessary to develop new, noninvasive muscle assessment techniques and to improve understanding of the relationship between muscle and functional capacity of elderly people. For the clinician and for epidemiological studies, a simple method is necessary to detect sarcopenia in large populations.

Recent work emphasizes current interest in anthropometric measurements and has suggested that calf circumference (CC) could be used to assess muscle mass.[2,8] CC has not been previously assessed as a screening tool for sarcopenia in the elderly. The three questions posed in this study are whether CC correlates with ASM, can be used as a measure of sarcopenia, and is related to physical function.

MATERIALS AND METHODS

From 1992 to 1994, 1,458 women aged 70 and older took part, in Toulouse, France, in the European Patient Information and Documentation Systems (EPIDOS) Study. EPIDOS is a prospective epidemiological study, conducted in five

From the *Internal Medicine Service and Gerontology Clinic (Professeur Albarède), Hôpital La Grave-Casselardit, Toulouse, France; †Exploration of Respiratory Function Service and Sports Medicine, Hôpital Purpan, Toulouse, France; and ‡Epidemiology and Community Health Laboratory, Faculté de Médecine, Toulouse, France.

Address correspondence to Dr. Yves Rolland, Service de Médecine Interne et de Gérontologie Clinique, Hôpital La Grave-Casselardit, 140 AV de Casselaudit, 31300 Toulouse, France. E-mail: yves.rolland@9online.fr

French towns (Amiens, Lyon, Montpellier, Paris, and Toulouse) to assess the risk factors of hip fracture in healthy elderly.[9]

Population and Protocol

The 1,458 women were recruited from electoral lists. Women were excluded if they were unable to walk independently, were living in an institution, had a previous history of hip fracture or hip replacement, or were unable to understand or answer the questionnaire. This study analyzed the data collected during the inclusion visit.

Anthropometric Measurement Methods

A trained technician using standardized techniques[10] in a research laboratory performed the anthropometric measurements. Weight was measured using a beam balance scale and height with a height gauge. Hip circumference was determined using a tape measure at the level of the maximum posterior protrusion of the buttocks. Waist circumference was measured 1 cm above the iliac crests. Grip strength was measured for the dominant hand with a hand dynamometer (Takei Ltd, Tokyo, Japan). The maximal value was recorded for a set of three contractions. CC was measured with the patient supine, with left knee raised and calf at right angles to the thigh. The tape measure was placed around the calf and moved to obtain the maximal circumference. Subcutaneous tissues were not compressed.

Muscular Mass Measurements

The reference method for the measurement of muscle mass was DEXA (QDR 4500 W, Hologic Inc., Bedford, MA). The DEXA was regularly calibrated. A trained technician performed the DEXA measurement protocol. Sarcopenia was defined according to the notion of ASM developed by Heymsfield.[11] ASM corresponds to the sum of the two arm and leg muscular masses in weight (kg).[11] A value relating to patient measurements is necessary to study ASM. Sarcopenia was then defined with ASM/height^2 or skeletal muscle mass (SMM) index in weight per meter^2 (kg/m^2).

This SMM index has been used previously to define sarcopenia.[2] Using this approach, participants with a SMM index of less than two standard deviations below the mean are considered sarcopenic. The reference population was a young female group from the Rosetta study and included adults aged 18 to 40.[12] This approach is the only one available to define a sarcopenic population and states that subjects with a SMM index lower than 5.45 for women and 7.26 for men are sarcopenic.

Health, Disability, and Physical Function Assessment

A physical examination and a health status questionnaire were used to record comorbid disease: hypertension, diabetes mellitus, dyslipidemia, coronary heart disease, peripheral vascular disease, cancer, stroke, Parkinson's disease, depression, and pain (of the back, hip, knee, ankle, or feet). Obesity was defined as a body mass index ((BMI) weight/height^2) above 30. Impaired vision was assessed using a visual acuity test. It was measured at a distance of 5 m with a Snellen letter test chart. Smoking (previous or current) and alcohol intake were noted. Monthly incomes were divided into three groups. Participants were asked about previous history of bone fracture, a fear of falling, and whether they had experienced one or more falls during the past 6 months.

Subjects answered a questionnaire about their ability to perform, without assistance, basic activities of daily living (ADLs) and instrumental activities of daily living (IADLs).[13,14] The IADL scale consists of eight items: food preparation, housekeeping, grocery shopping, doing laundry, handling money, using the telephone, taking medications, and using public transportation. Only those IADL variables that involved locomotion (food preparation, housekeeping, grocery shopping, doing laundry, and using public transportation) were used.

Participants were also asked whether they had difficulties (no/some/serious difficulty) performing various physical movements, such as walking, climbing stairs, rising from a chair or bed, picking up an object from the floor, lifting heavy objects, or reaching an object. Women with some or serious difficulties were grouped together. Women with three or more difficulties (labeled as "moving difficulties") were also grouped together.

Women were asked whether they had taken hormone replacement therapy or corticosteroids during at least 3 months. They also reported whether they regularly did sport activities or physical activities such as hiking, gymnastics, cycling, swimming, or gardening.

Statistical Methods

Analysis of the relationship between anthropometric factors and the muscular mass measured by DEXA was performed with a bivariate analysis, using the correlation coefficient for the associations between the quantitative variables and the Student t tests for mean comparisons. To consider the confounding, a multiple linear regression was realized. The initial model consisted of variables found to be associated with ASM to a threshold of 0.25 in bivariate analysis. A step-by-step decreasing regression was done to obtain the best reduced model.

The most discriminating CC value that separated the sarcopenic from the nonsarcopenic subjects (defined using DEXA) was determined. A receiver operating characteristic (ROC) curve was used to ascertain the discriminatory value for CC. These curves explained the performance of each CC to separate the subjects with or without sarcopenia (defined using DEXA).

Associations between sarcopenia, disability, physical function, and risk of fall were assessed. Several logistic models were used. Variables that needed to be explained were disability, physical function, and risk of fall. The explicative variables were sarcopenia defined using the SMM index measured by DEXA or by the CC. Age, obesity, income, hypertension, diabetes mellitus, dyslipidemia, coronary heart disease, peripheral vascular disease, cancer, stroke, Parkinson's disease, depression, pain, visual impairment, corticosteroid treatment, smoking, and alcohol intake were adjusted.

RESULTS

The EPIDOS (Toulouse) population included 1,458 women. The ASM and CC were measured in 90% of the population. The final sample included 1,311 women with a mean age ± standard deviation of 80.3 ± 3.8. Patients aged 75 to 85 represented 84% of the population. One hundred

Table 1. Anthropological Values and Muscle Mass Measured Using Dual-Energy X-Ray Absorptiometry

Variable	Mean ± Standard Deviation	Minimum	Maximum
Weight, kg	58.7 ± 9.7	33	96
Height, cm	152.9 ± 5.9	130	173
Body mass index, kg/m^2	25.1 ± 3.9	15.1	40.4
Waist circumference, cm	85.1 ± 9.9	59	124
Hip circumference, cm	99.4 ± 8.6	79	134
Calf circumference, cm	35.0 ± 3.1	25	48
Upper limb strength, kPa	54.3 ± 13.2	12	105
Upper limb muscle mass, kg	3.9 ± 0.8	1.8	8.3
Lower limb muscle mass, kg	11.0 ± 1.4	6.9	15.9
Appendicular muscle mass, kg	14.9 ± 2.0	9.3	22.8
Skeletal muscle mass index, kg/m^2	6.4 ± 0.8	4.5	10.0

twenty-four women had sarcopenia, a prevalence of 9.5% (95% confidence interval = 7.9–11.1). Association of anthropometric factors with ASM was measured using DEXA.

Table 1 shows the anthropometric data for the study population. The mean CC was 35 ± 3.0 cm (range 25–48 cm). The mean ASM was 14.9 ± 2.0 kg.

Table 2 shows the correlations between the different anthropometric variables and ASM. In this study, weight, height, waist circumference, hip circumference, CC, BMI, and grip strength were positively correlated with ASM. Age was negatively correlated with ASM. Weight was the most strongly correlated variable with ASM ($r = 0.73$). Of the anthropometric measurements, CC was most strongly correlated with ASM ($r = 0.63$). There was no statistically significant difference in ASM between subjects with history of stroke (n = 7), corticosteroid treatment (n = 59), hormone replacement therapy (n = 221) and those who self-reported usual sport activities (n = 497).

The best predictive equation for the ASM developed using stepwise regression using data for all the subjects was ASM (kg) = 0.31 CC (cm) + 0.05 waist circumference (cm) + 0.08 hip circumference (cm) + 0.02 grip strength (kPa) − 0.16 BMI (kg.m^{-2}) − 5.1.

Forty-eight percent of the variance in ASM can be predicted from the variables CC, waist circumference, hip circumference, grip strength, and BMI.

Best Discriminatory CC for Determining Patients with or without Sarcopenia

From the ROC curve, the discriminatory CC value, statistically defined as the best compromise between sensitivity and specificity, was 32 cm. A CC of 31 cm was chosen to maximize specificity and therefore positive predictive value for sarcopenia without reducing sensitivity too much. ROC curves including other anthropometric variables such as weight or height did not increase the area under the curve or improve the efficiency of the test. The sensitivity and specificity of CC lower than 31 cm for the diagnosis of sarcopenia was, respectively, 44.3% and 91.4%. The positive predictive value was 35.1%.

Association Between Sarcopenia, Disability, Physical Function, and Risk of Fall

Table 3 shows the association between sarcopenia defined by a SMM index of less than 5.45 or CC of 31 cm or less and markers of disability, physical function, and risk of fall. Women with a CC of 31 cm or less were 2.6 times more likely to have difficulties with bathing/dressing or walking. A small CC value (≤31 cm) was twice as often associated with difficulties in lifting a heavy object, and more than three times more often associated with difficulty in three or more of the following activities: walking, going upstairs or downstairs, standing up from a chair, picking something from the floor, lifting a heavy object, reaching an object, and rising from bed. In this group, there was no increased risk of fall.

In this population from Toulouse, women with sarcopenia defined by SMM index were not significantly more likely to experience disability, difficulty with physical function, or risk of fall than nonsarcopenic women.

DISCUSSION

The first aim of the EPIDOS Study was to investigate hip fracture risk factors.[16] This work is a cross-sectional analysis of a selected elderly population. Patients included in this study were all volunteers living at home with no difficulties in walking alone and no history of hip fracture or hip replacement. This selection of a healthy population is not representative of the general population of the same age.

The muscle mass of this elderly population was lower than of young subjects as reported by numerous previous studies.[2,12] The selection of a healthy population with walking autonomy explains the differences observed in other elderly populations even if an ethnic or cultural factor can not be excluded, because the population of the New Mexico Elder Health Study also had a healthy, ambulatory population. Women from the Aging Process Study[4] or the New Mexico Elder Health Survey[2] were on average younger, heavier, and taller than the women from Toulouse, but the ASM of the subjects, measured using the same methodology, was higher. The low prevalence of sarcopenia (9.5%)

Table 2. Correlation Between Anthropometric Variables and Appendicular Muscle Mass

Variable	Correlation Coefficient	95% Confidence Interval
Age	−0.13	−0.2 to −0.08
Weight	0.73	0.71–0.76
Height	0.49	0.45–0.53
Body mass index, kg/m^2	0.53	0.49–0.57
Waist circumference	0.51	0.47–0.55
Hip circumference	0.58	0.55–0.62
Calf circumference	0.63	0.60–0.66
Grip strength	0.24	0.19–0.29

Note: $P < .001$ for all correlations.

Table 3. Adjusted* Association Between Sarcopenia† and Disability, Self-Reported Physical Function, and Risk of Fall

Disability	%	Sarcopenia Defined by DEXA with SMM Index <5.45			Sarcopenia Defined by CC <31 cm		
		OR	95% CI	P-value	OR	95% CI	P-value
Activity of daily living disabilities							
Walking difficulties	6.8	1.47	0.65–3.32	.36	2.67	1.36–5.24	.004
Bathing/dressing or walking difficulties	7.2	1.55	0.71–3.37	.27	2.62	1.37–5.04	.003
IADL disabilities							
IADL 8 items‡ (≥3 incapacities)	3.4	0.74	0.22–2.52	.63	1.78	0.74–4.33	.195
IADL 5 items§ (≥3 incapacities)	2.7	0.69	0.16–2.89	.61	2.49	0.94–6.63	.067
Physical function difficulties							
Ascending stairs	7.8	1.51	0.73–3.13	.27	1.94	1.03–3.65	.041
Descending stairs	6.2	1.06	0.46–2.47	.89	2.15	1.11–4.15	.022
Rising from a chair	2.8	0.64	0.13–3.05	.57	2.15	0.77–5.99	.144
Picking an object from the floor	4.7	0.77	0.27–2.23	.63	1.38	0.62–3.06	.424
Lifting heavy objects	11.8	1.68	0.92–3.05	.09	2.45	1.45–4.12	.001
Moving difficulties‖ (≥3 difficulties)	6.0	1.41	0.59–3.36	.44	3.19	1.61–6.32	.001
Risk of fall							
Fear of falling	58.0	0.98	0.65–1.47	.91	1.21	0.84–1.75	.303
Fall during the 6 previous months	23.5	1.02	0.66–1.59	.92	1.11	0.75–1.64	.613
History of bone fracture	41.7	0.75	0.51–1.12	.15	0.75	0.52–1.07	.114

* Adjusted for age, obesity, income, hypertension, diabetes, dyslipidemia, coronary heart disease, peripheral vascular disease, cancer, stroke, Parkinson's disease, depression, pain, visual impairment, steroid treatment, smoking, and alcohol intakes.
† Appendicular muscle mass/height^2 (kg/m^2) less than two standard deviations below the mean value for young women from the Rosetta Study.[19] Defined using dual-energy x-ray absorptiometry (DEXA) and calf circumference (CC) <31 cm.
‡ Instrumental activity of daily living (IADL) 8 items: food preparation, housekeeping, shopping for groceries, doing laundry, handling money, using the telephone, taking medications, using public transport.
§ IADL 5 items: food preparation, housekeeping, shopping for groceries, doing laundry, using public transport.
‖ Moving difficulties: walking, climbing stairs, rising from a chair or bed, picking an object from the floor, lifting heavy objects, reaching an object.

in this sample probably reflects the ability of this population to ambulate.[2,4] This study confirms findings from another study suggesting that sarcopenia affects 6% to 15% of persons aged 65 and older.[15]

The average CC of the patients in this study was higher than that for elderly suffering from undernutrition (CC = 29.2 ± 3.6 cm)[8] or geriatric inpatients (CC = 29.9 ± 4.3 cm).[6] CC therefore seems to have clinical associations.

CC is an anthropometric variable correlated with ASM, but this correlation is low ($r = 0.63$), and explains only 40% of the variance. Measurement error can be envisaged but a trained technician precisely measured CC and ASM. The weakness of the relation between ASM and CC may be due to the good health of this population. Anthropometric studies report a higher correlation between CC and total muscle mass in malnourished ($r = 0.67$)[6] or dead subjects ($r = 0.91$).[16] Subcutaneous fat, which is often minimal in ill, malnourished, or dying elderly patients, is a confounding factor. Therefore, in this situation, CC is more representative of muscle mass. The opposite is observed in obese subjects.[8,17]

This study did not assess arm circumference, which has been used in young populations to estimate total muscle mass.[18] CC seems nevertheless to provide a better estimation of muscle mass in the elderly.[19] In the New Mexico Aging Process Study,[4] the correlation between CC and ASM was twice the correlation between arm circumference and ASM. Hip and waist circumferences and BMI give information concerning fat mass and cardiovascular risks. Sarcopenic patients had a BMI significantly lower than nonsarcopenic subjects. The correlation is probably due to the relationship between muscle mass and nutritional status.[8] The few studies estimating muscle mass from anthropometric data are not convincing.[16,20] By explaining only 48% of the variance, the best equation fails to predict the ASM mass. Of the anthropometric variables, CC is correlated with ASM, but its low sensitivity makes it a poor screening tool for sarcopenia as defined by DEXA.

Associations between muscle mass and strength or between strength and functional performance are known, but only one study reported a direct link between low muscle mass and functional incapacity.[2] In this study, sarcopenic women (defined by DEXA) had four times more chance of having difficulties performing three or more IADLs. This result confirms the theory concerning the key role of muscle loss in the development of disability. Despite similar methodology, this association was not observed in this study. Sarcopenic women (defined by DEXA) have no more disability or greater self-reported difficulties for physical function. This result casts doubt on the validity of the chosen criterion for sarcopenia. The amount of muscle mass lost in sarcopenic patients is still unknown. The threshold chosen to define sarcopenia by means of DEXA is arbitrary.[2] The lack of association between sarcopenia and disability in this population suggests that a lower threshold should be adopted. In addition, the young U.S. female reference group is perhaps not the best to develop the criteria of sarcopenia for a French population.

This result also casts doubt on the purely quantitative definition of sarcopenia used in this study. The DEXA measurement used in this study did not measure the quality of the muscle. Sarcopenia is defined not only by loss of muscle quantity, but also by loss of muscle quality.[21] Whether functional capacity loss in elderly people is mainly due to muscle mass or qualitative impairment is unknown.[22]

Women with lower CC have greater disability and lower physical function. This cross-sectional study did not allow for the establishment of the relationship of cause and effect between CC, loss of autonomy, and physical function. It seems that cross-sectional area is better related to muscle function than ASM. ASM by DEXA is a measure of lean mass, which estimates muscle volume, but CC estimates cross-sectional muscle area and skin fold. It is likely that this cross-sectional area predicts factors (other than sarcopenia) associated with autonomy loss. Pain, depression, and malnutrition may also cause impaired physical function.[23] In general, ill people have lower CCs than healthy people.[7] Patients with low CC often present with poor nutritional status.[6] CC atrophy coexists with mobility impairment or functional incapacity, but future research needs to determine whether atrophy of muscle happens first, followed by impairment, or whether impairment leads to muscle atrophy.

In addition, these results confirm the latest studies, which assign a major role to leg physical function, particularly to the sural triceps,[24] with the maintenance of autonomy.[25-27] Women with a CC of less than 31 cm may have leg weakness.[25,27] Minimal leg strength is required for most daily activities. In this study, women presenting with a CC of less than 31 cm were three times more likely to have difficulties moving (Table 3). By contrast, arm strength did not show up in the literature as a predictive factor of functional decline.[26] This study therefore suggests that topography of muscle loss is more associated than global muscle mass loss with physical function and disability.

This study is the only large epidemiological study of the relationship between anthropometric variables, muscle mass, and self-reported physical function. Of the study variables, CC is not a good screening tool for sarcopenia (defined using DEXA). By contrast, CC is a potential marker of physical function. It is a cheap, simple, noninvasive measurement for a clinician and seems relevant in the screening of sarcopenia defined as "a condition in which muscle strength is insufficient to perform the normal tasks associated with an independent lifestyle."[28]

REFERENCES

1. Rosenberg IH. Summary comments. Am J Clin Nutr 1989;50:1231–1233.
2. Baumgartner RN, Koehler KM, Gallagher D et al. Epidemiology of sarcopenia among the elderly in New Mexico. Am J Epidemiol 1998;147:755–763.
3. Carmeli E, Reznick A. The physiology and biochemistry of skeletal muscle atrophy as a function of age. Proc Soc Exp Biol Med 1994;206:103–113.
4. Baumgartner RN, Stauber PM, McHugh D et al. Cross-sectional age differences in body composition in persons 60+ years of age. J Gerontol A Biol Sci Med Sci 1995;50A:M307–M316.
5. Poehlman ET, Toth MJ, Gardner AW. Changes in energy balance and body composition at menopause: A controlled longitudinal study. Ann Intern Med 1995;123:673–675.
6. Jauffret M, Jusot JF, Bonnefoy M. Marqueurs anthropométriques et malnutrition de la personne âgée: intérêt de la circonférence du mollet. Age Nutr 1999;10:163–169.
7. Heymsfield SB, Gallagher D, Visser M et al. Measurement of skeletal muscle: Laboratory and epidemiological methods. J Gerontol A Biol Sci Med Sci 1995;50A Spec No:23–29.
8. Chumlea WC, Guo SS, Vellas B et al. Techniques of assessing muscle mass and function (sarcopenia) for epidemiological of the elderly. J Gerontol A Biol Sci Med Sci 1995;50A:45–51.
9. Dargent-Molina P, Favier F, Grandjean H et al. Fall-related factors and risk of hip fracture: The EPIDOS prospective study. Lancet 1996;348:145–149.
10. Lohman TG, Roche AF, Martorell R, eds. Anthropometric standardization reference manual. Champaign, IL: Human Kinetics, Inc., 1988.
11. Heymsfield SB, Smith R, Aulet M et al. Appendicular skeletal muscle mass: Measurement by dual-photon absorptiometry. Am J Clin Nutr 1990;52:214–218.
12. Gallagher D, Visser M, De Meersman RE et al. Appendicular skeletal muscle mass. Effects of age, gender, and ethnicity. J Appl Physiol 1997;83:229–239.
13. Katz S, Downs TD, Cash HR et al. Progress in development of the index of ADL. Gerontologist 1970;10:20–30.
14. Lawton MP, Brody EM. Assessment of older people: Self-maintaining and instrumental activities of daily living. Gerontologist 1969;9:179–186.
15. Melton LJ, Khosla S, Crowson CS et al. Epidemiology of sarcopenia. J Am Geriatr Soc 2000;48:625–630.
16. Martin AD, Spenst LF, Drinkwater DT et al. Anthropometric estimation of muscle mass in men. Med Sci Sports Exerc 1990;22:729–733.
17. Heymsfield SB, Olafson RP, Kutner MH et al. A radiographic method of quantifying protein-calorie undernutrition. Am J Clin Nutr 1979;32:693–702.
18. Friedman PJ, Campbell AJ, Caradoc-Davies TH. Prospective trial of a new diagnostic criterion for severe wasting malnutrition in the elderly. Age Ageing 1985;14:149–154.
19. Patrick JM, Bassey EJ, Fentem PH. Changes in body fat and muscle in manual workers at and after retirement. Eur J Appl Physiol 1982;49:187–196.
20. Heymsfield SB, McManus C, Smith J et al. Anthropometric measurement of muscle mass: Revised equations for calculating bone-free arm muscle area. Am J Clin Nutr 1982;36:680–690.
21. Dutta C, Hadley E, Lexel J. Sarcopenia and physical performance in old age. Overview. Muscle Nerve 1997;5:S5–S9.
22. Schwartz R. Sarcopenia and physical performance in old age: introduction. Muscle Nerve 1997;5:S10–S12.
23. Brown M, Sinacore D, Host HH. The relationship of strength to function in the older adult. J Gerontol 1995;50 Spec No:55–59.
24. Bassey EJ, Bendall MJ, Pearson M. Muscle strength in the triceps surae and objectively measured customary walking activity in men and women over 65 years of age. Clin Sci 1988;74:85–89.
25. Guralnik JM, Ferrucci L, Pieper C et al. Lower-extremity function and subsequent disability: Consistency across studies, predictive models, and value of gait speed alone compared with the short physical performance battery. J Gerontol A Biol Sci Med Sci 2000;55A:M221–M231.
26. Schenkman M, Hughes MA, Samsa G et al. The relative importance of strength and balance in chair rise by functionally impaired older individuals. J Am Geriatr Soc 1996;44:1441–1446.
27. Ferrucci L, Guralnik J, Buchner D et al. Departure from linearity in the relationship between measures of muscular strength and physical performance of the lower extremities: The women's health and aging study. J Gerontol A Biol Sci Med Sci 1997;52A:M275–M285.
28. Bassey EJ. Measurement of muscle strength and power. Muscle Nerve 1997;5:S44–S46.

"Opinion formation in evaluating sanity at the time of the offense: An examination of 5175 pre-trial evaluations," J. I. Warren, et al, *Behavioral Science and the Law*, 22:171–186, 2004.[1]

On p. 177, in the last paragraph, the authors mention DSM-III-R and DSM-IV. DSM stands for Diagnostic and Statistical Manual, which is used to diagnose illness (in this case, mental illness).

1. Consider Table 1. How was the value of 21.24 calculated? It has several asterisks after it. What does this mean? How would you give the meaning for 21.24 in the form a sentence that addresses the issues of the paper?
2. Table 1 contains CC = contingency coefficient. You might use the Internet to learn more about this quantity. It is said to be "a measure of effect size." How does this differ from the quantity in the next to last quantity?
3. Table 3 contains 7 entries under "Type of crime." A statistic is computed for each type. Why didn't they compute a single statistic for this category? What would such a statistic mean?
4. If we add the four counts for Age, we get 1673. If we add the four counts for Gender, we get 3822. Should these two totals be the same? Why aren't they?
5. Compare and contrast Table 2 and Table 1 in terms of what they tell us about the formation of opinion.
6. Under Psychiatric diagnoses, there are some very large values in the next to last column. What do these indicate? How would the percentages for these items compare to, say, the percentages for Gender?
7. Table 3 compares psychologists and psychiatrists. The first item is Opinion. What does Table 3 say about the rates at which psychologists and psychiatrists find defendants sane or insane? How would you find a confidence interval for the difference between the two groups? What would the interval tell you that the table does not?
8. Do psychologists or psychiatrists have a higher rate of finding the defendant insane? Does this tell you about psychologists and psychiatrists or the type of subjects that were examined by each? Were the subjects assigned randomly to psychologists and psychiatrists? How would that have mattered?
9. Table 4 shows the results of a logistic regression model. You might not have covered logistic regression. It is a model using several variables (in the left hand column of Table 4) to predict the probability that the evaluator will return an opinion of insane. The second column contains the coefficients of the model. Beside it are the standard errors. What do you think is the distribution (approximately) of the coefficients? Based on this, check the *P*-values. (Bear in mind that we don't have all the decimal places for the coefficient and standard error, so our calculations might be off a little.)

---

[1] *© 2003, John Wiley & Sons, Ltd. Used by permission.*

Opinion Formation in Evaluating Sanity at the Time of the Offense: An Examination of 5175 Pre-Trial Evaluations

Janet I. Warren, D.S.W*, Daniel C. Murrie, Ph.D., Preeti Chauhan, B.S., Park E. Dietz, M.D., Ph.D., M.P.H. and James Morris, Ph.D.

Sanity evaluations are high-stake undertakings that explicitly examine the defendant's culpability for a crime and implicitly explore clinical information that might inform a plea agreement. Despite the gravity of such evaluations, relatively little research has investigated the process by which evaluators form their psycholegal opinions. In the current study, we explore this process by examining 5175 sanity evaluations conducted by a cohort of forensic evaluators in Virginia over a ten-year period. Our analyses focus on (i) the clinical, criminal, and demographic attributes of the defendant correlated with opinions indicative of insanity; (ii) the clinical content of the evaluations and the legal criteria referenced as the basis for the psycholegal opinion; (iii) the process and outcome differences in the sanity evaluations conducted by psychologists versus psychiatrists; and (iv) the consistency in these opinions over a ten year period. Analyses predicting an opinion of insanity indicate a positive relationship with psychotic, organic, and affective diagnoses and previous psychiatric treatment. Analyses also indicate a negative relationship with prior criminal history, drug charges, personality disorder diagnosis, and intoxication at the time of the offense. Modest racial disparities were observed with evaluators offering opinions that the defendant was insane more often for white than for minority defendants despite comparable psychiatric and criminal characteristics. Copyright © 2003 John Wiley & Sons, Ltd.

---

*Correspondence to: Janet I. Warren, Institute for Law, Psychiatry, and Public Policy, University of Virginia Health System, P.O. Box 800660, Charlottesville, VA 22908-0660. U.S.A.
E-mail: jiw@virginia.edu

This research was funded under contract by the Virginia Department of Mental Health, Mental Retardation and Substance Abuse Services and the Office of the Attorney General of the Commonwealth of Virginia. The points of view expressed in this document are those of the authors and do not necessarily represent the official position of the Virginia Department of Mental Health, Mental Retardation and Substance Abuse Services or the Office of the Attorney General.

Copyright © 2003 John Wiley & Sons, Ltd.

INTRODUCTION

Although the field of forensic psychology has demonstrated substantial expansion and maturation during its brief 50 year history, practice guidelines remain ill defined for many types of forensic evaluation (Otto & Heilbrun, 2002). Indeed, a review of the research reveals such variability in clinical–forensic practice that some scholars have concluded that the level of forensic practice "falls far short of professional aspirations for the field" (Nicholson & Norwood, 2000, p. 9). Such a status warrants concern, given the trend for both federal and state courts to require an increasingly empirical standard of practice (*Daubert v. Merrell Dow Pharmaceuticals*, 1993; *Kumho Tire Co. v. Carmichael*, 1999) and growing societal concerns about the relationship between psychiatric disorders and violent crime.

How, then, is forensic practice conducted? What exactly do evaluators *do* to arrive at their psycholegal opinion? The research addressing the process or outcome of forensic mental health evaluations has tended to focus on trial competence. Traditionally, these studies have examined the factors associated with a finding of incompetence (Cooper & Grisso, 1997; Grisso, 1992; Hart & Hare, 1992; Nicholson & Kugler, 1991; Rosenfeld & Wall, 1998). More recently, studies addressing competence evaluation have begun to examine the reliability among evaluating clinicians (Rosenfeld & Ritchie, 1998) as well as the factors considered and the process documented (Nicholson & Norwood, 2000; Skeem, Golding, Cohn, & Berge, 1998). Other emergent research explores the roles that psychological testing (Borum & Grisso, 1995; Heilbrun, 1990), third party information (Heilbrun & Collins, 1995; Heilbrun, Rosenfeld, Warren, & Collins, 1994; Heilbrun, Warren, & Picarello, 2003), and different methods of quantification (Slovic, Monahan, & MacGregor, 2000) can play in forensic evaluation and opinion formation.

Less often available in forensic research has been the empirical study of sanity at the time of the offense, even though these pre-trial evaluations are central not only in determining a defendant's culpability for often serious felony crimes, but also in discerning clinical information that might be relevant to plea bargaining and sentencing. This relative lack of empirical research may reflect the smaller number of sanity evaluations conducted each year and the low base rate for an insanity finding by evaluators or the courts (Warren, Fitch, Dietz, & Rosenfeld, 1991). Undoubtedly, however, it also reflects the complex nature of the sanity evaluation and the difficulties intrinsic to studying it empirically.[1] Unlike the relatively consistent battery of questions used in standard evaluations of trial competence, sanity evaluations require a retrospective examination of mental state, which demands a detailed scrutiny of intrinsically unique crimes, as well as integration of clinical and collateral information. To the extent that subjective clinical impressions guide the inquiry and determine the progression from clinical observation to psycholegal opinion formation, the process may be too covert to permit direct observation of the sanity evaluation and its conclusions.

Given the limits of the available research, we attempt to describe these dynamics by examining the process used and the conclusions reached in 5175[2] sanity

---

[1] For a more on conducting sanity evaluations and empirical research using a standardized measure designed to address legal sanity, see Rogers and Shuman (2000).

[2] Although the overall $N = 5175$, ns vary by analysis. In some cases, certain variables were missing, while in other cases we created a smaller subset ($n = 3174$) to control for the effects of a few prolific evaluators.

evaluations conducted in Virginia over a 10 year period. In particular, we focus on (i) the clinical, criminal, and demographic attributes of the defendant that correlate with opinions indicative of insanity; (ii) the forensic process and legal criteria used by the evaluating clinician(s) in reaching their psycholegal opinions; (iii) the process and outcome differences in the sanity evaluations conducted by psychologists versus psychiatrists; and (iv) the consistency in these opinions reached by a large cohort of evaluators over a 10 year period. These analyses are intended to further describe the reliability and validity of these sanity evaluations as well as to inform debate over the relative ability of different mental health professionals to perform these evaluations.

Offenses, Diagnoses, and Psycholegal Opinion

A few studies of large samples have investigated the relationship between criminal offenses or psychiatric diagnoses and subsequent psycholegal opinion regarding sanity at the time of the offense. Warren and colleagues (1991) examined 894 pretrial referrals in Virginia, 617 of which featured sanity as a referral issue either alone or in addition to competency to stand trial (CST). In that study, evaluators reached an opinion supporting an insanity claim in only 47 (8%) of the cases. Defendants' diagnoses were significantly related to the legal opinion. Not surprisingly, schizophrenia (13 defendants, 28%) was the most frequently cited diagnostic category among the 47 defendants opined to be insane. Affective disorder (15% of those opined insane) and mental retardation (11% of those opined insane) were also relatively common diagnoses given in evaluations that supported a finding of insanity. In contrast, diagnoses of personality disorders and substance abuse disorders were least likely to be associated with a finding of insanity. The above findings are consistent with the conclusions that Melton, Petrila, Poythress, and Slobogin (1997) drew after reviewing six smaller studies conducted across four states at different points between 1967 and 1987. Namely, the authors reported, "data suggests that the presence of a major psychosis is required for the insanity defense to succeed," particularly in more recent years (p. 216). Similarly, Callahan, Steadman, McGreevy, and Clark-Robins (1991) reported that 55% of those who pled not guilty by reason of insanity and 84% of those acquitted as such carried a diagnosis of "schizophrenia or another major mental illness (other psychotic or affective disorder)" (p. 336).[3]

In the Warren et al. (1991) study, the category of offense was also significantly related to the opinion supporting insanity. For example, between 0 and 4% of defendants charged with robbery, sex offending, or murder were opined to be insane, as compared with 10% of defendants charged with property crimes and public order offenses.

A decade later, Cochrane, Grisso, and Frederick (2001) conducted a similar study, examining competence and sanity opinions among 1710 federal pretrial defendants, and found similar results. Their data reflected a strong relationship between diagnoses and psycholegal opinion: 40% of defendants with psychotic

---

[3]The stated goal of the Callahan et al. (1991) study was to present a systemic description of "the volume, rates, and composition of insanity pleas and acquittals across states" (p. 332). As such, it differs from the other studies reviewed, and the present one, which report the evaluator's clinical or psycholegal opinion rather than a legal strategy (entering an insanity plea) or judicial decision (acquittal on such a plea).

disorders were opined to be insane as compared to only 6% of those with personality disorders. Regarding criminal charges, insanity rates varied greatly based on the type of offense. For example, those charged with threatening a government official or assault were most likely to be found insane by forensic examiners (36 and 31%, respectively), as compared with none of the defendants charged with sex crimes or kidnapping.

Unlike previous research, Cochrane and colleagues investigated the relationship between diagnoses and charges. The authors proposed that the high rates of insanity for certain crime categories were best explained by the high rates of psychotic diagnoses for defendants within these crime categories. Consistent with their hypothesis, logistic regression revealed that there were no significant relationships between charges and psycholegal opinion once diagnoses were also considered. Rather, diagnoses were related to the types of offense the defendant committed and diagnostic presentation was the main variable to affect psycholegal opinion. This conclusion about the primacy of diagnostic presentation is particularly noteworthy when compared with a study of trial competence (e.g. Rosenfeld & Ritchie, 1998) that demonstrated a relationship between charges and psycholegal opinion and concluded that forensic evaluators may be biased towards particular conclusions due to the severity of the offense.

Other Factors Involved in Psycholegal Opinion Formation

Beyond the defendant's diagnostic presentation, what other factors do examiners consider in sanity evaluations? Interestingly, the forensic literature appears to offer more guidance regarding what evaluators *should* consider rather than what they *do* consider in practice. For example, several authors have provided well reasoned guidelines about the ways in which psychological testing (see, e.g., Borum & Grisso, 1996; Heilbrun, 1992) and third party information (Heilbrun et al., 1994, 2003) should inform forensic evaluations. However, normative data regarding the degree to which evaluators actually use such sources of clinical information is more sparse. Among other benefits, such normative data could be useful in evaluating some of the critiques of forensic evaluations (Heilbrun & Collins, 1995). For example, Grisso's (1986) summary of the criticisms levied against forensic practice included the claim that evaluators failed to collect sufficient or legally relevant information on which to base their opinion.

An early study by Petrella and Poythress (1983) yielded data regarding the sources of information forensic evaluators consider when performing sanity evaluations. For the 80 sanity cases examined, 33% of reports documented consultation with the defendant's attorney, 26% documented a review of previous medical or psychiatric records, and 14% documented some type of consultation with other professionals.

Heilbrun and Collins (1995) also examined forensic evaluation reports ($n = 277$) including evaluations performed in the community ($n = 110$) as well as those from an inpatient state hospital ($n = 167$) to investigate differences between settings. The legal issues addressed in this sample of reports included competence, sanity, both, or other. For cases in which sanity was at issue (either alone or in addition to competence), the authors presented information only on procedures used by the community sample. A clinical interview was used in 98–100% of cases, a mental status exam was performed in 67–69% of cases, an arrest report was reviewed in 65–67% of cases, and prior mental health records were reviewed in only 31–33% of the cases.

Regarding the use of psychological testing in criminal forensic evaluation, only 16% of the reports in the Heilbrun and Collins (1995) sample (which included hospital and community evaluation of competence, sanity, and other issues) mentioned testing, with the MMPI and WAIS-R cited most frequently. Testing was used more often in hospital-based reports, but this is likely a function of the evaluators' discipline; hospital-based evaluators were psychologists, whereas most community-based evaluators were psychiatrists. A full breakdown of test usage by legal issue (allowing for examination of test administration in all cases where sanity was raised as an issue) was not available. Finally, Borum and Grisso (1995) surveyed 53 psychologists and 43 psychiatrists who conducted Criminal Responsibility (sanity) evaluations. Among forensic psychologists, 68% reported that they used testing "frequently" or "always," whereas 32% reported that they used testing only "sometimes" or "rarely." Rates were slightly lower among forensic psychiatrists, of whom 42% reported that they "frequently" or "always" use psychological testing.

Interdisciplinary Differences

Another focus in the research has examined whether forensic evaluators of different disciplines differ in the degree to which they use various sources of information. Although the Heilbrun and Collins (1995) study included variables reflecting the discipline to which an evaluator belonged, the authors did not perform analyses contrasting the procedures employed by psychiatrists versus psychologists. However, the Petrella and Poythress (1983) study was designed primarily to compare the quality of forensic evaluations between disciplines. The authors examined both the final written product and the sources of information consulted during the evaluation process. For sanity evaluations, psychologists sought outside sources of information more frequently than did psychiatrists (0.88 versus 0.58 average consultations per case), and produced a greater quantity of "clinical notes" (172 versus 136 average lines of clinical notes per case), both of which were interpreted as suggesting a more thorough evaluation. When two legal experts, a trial judge and a prosecuting attorney, rated the quality of the final reports, those by psychologists tended to be scored more favorably, with differences reaching statistical significance on two criteria. Results from this study "fail(ed) to support the conventional wisdom that psychiatrists perform forensic evaluations that are superior to those conducted by non-medically trained clinicians" (Petrella & Poythress, 1983, p. 76).

Despite two decades of evolving forensic practice and the findings from the Petrella and Poythress (1983) study, the same "conventional wisdom" (p. 76) those authors described apparently prevails today. A recent study (Redding, Floyd, & Hawk, 2001) surveyed Virginia trial judges ($n=59$), prosecutors ($n=46$), and defense attorneys ($n=26$) regarding their preferences for various aspects of mental health testimony. When asked to rank which mental health professional they would prefer to have complete a forensic psychological evaluation for the court, the respondents favored a psychiatrist, followed by a PhD psychologist, followed by other clinicians. Sixty-eight percent of respondents ranked a psychiatrist first in preference, while only 32% ranked a PhD psychologist first, indicating a significant preference for psychiatrists ($\chi^2(1) = 24$, $p < 0.001$). Given these preferences, it would be worthwhile to investigate in a large sample what differences, if any, actually exist between psychologists and psychiatrists in the practice of forensic evaluation.

Current Study Objectives

In the current study, we investigate these various considerations by examining the content and process of 5175 evaluations conducted in Virginia by a cohort of psychiatrists and clinical psychologists over a 10 year period. Analyses examine factors that were related to a psycholegal opinion regarding insanity, the process used by psychologists and psychiatrists in reaching their conclusions, disciplinary differences in opinion formation, and the consistency and change in these opinions over a 10 year period. We also report the cognitive and volitional prongs of the insanity defense the evaluators identify in their opinions supportive of insanity.

METHOD

Participants

We reviewed the characteristics of 5175 sanity at the time of offense evaluations conducted by evaluators trained by the Institute of Law, Psychiatry and Public Policy at the University of Virginia over a 10 year period. Each of the evaluators had completed a 7 or 5 day training program that addressed both the clinical and legal parameters of evaluating legal sanity and the unique demands of this particular type of evaluative setting. Participants in the training program were required to submit an opinion segment of a report for review and to pass an examination at the end of the training program.

Measures

Data were compiled using the Forensic Evaluation Information Form, a 2 page instrument that includes information concerning (i) the defendant's prior criminal and psychiatric histories, compliance at the time of the offense with prescribed psychotropic medication, use of alcohol or other non-prescribed substances at the time of the offense, and current and past psychiatric condition and criminal history; (ii) the nature of the examination, including the number and discipline of the evaluators, the time spent on different components of the evaluation, the use of psychological tests, and the sources of information both requested and obtained; and (iii) the nature of the psycholegal opinion in terms of both a penultimate opinion and the prongs of the insanity defense that are identified as suggesting impairment or not. The Forensic Evaluation Information Form is submitted voluntarily by evaluators practicing privately as well as those based in hospitals, community service boards, and other institutions. Earlier research suggests that 50–80% of court-ordered forensic evaluations conducted in Virginia each year are entered into the Forensic Evaluation Information System (FEIS; Warren, 1992).

Evaluators

A total of 222 evaluators (164 psychologists and 58 psychiatrists) evaluated 5073 cases. In 102 additional cases the individual evaluator was unknown; however the discipline of the evaluator was known, allowing the cases to remain in the dataset for analyses. Two psychologists evaluated 32% of all cases evaluated by their discipline. One evaluated 856 cases and found 34 (4%) of these to be insane; another evaluated 603 cases and found 26 (4%) of these to be insane. Similarly, two psychiatrists evaluated 50% of the cases conducted by psychiatrists. One evaluated 118 cases and found 11 (9%) to be insane and the other evaluated 210 cases and found 22 (10%) to be insane. To avoid having these four evaluators unnecessarily skew the results of the comparisons between the two disciplines, only the mean number of evaluations from each discipline ($n=27$ for psychologists, $n=11$ for psychiatrists) were randomly selected and allowed to remain in the dataset for these four evaluators. However, this smaller subset ($n=3471$) was only used for analyses examining interdisciplinary differences and not for those analyses examining opinion variables overall.

Age

Subtracting the date of birth from the date of the evaluation yielded age. Only defendants who were 18 or older at the time of the evaluation were used in age related data analyses. The mean age (36) was used to divide the variable into older or younger defendants.

Offense

The instant offense was coded using the seven-category classification system developed by Policy Research Associates, Inc. for use in cross state studies. These include: violent, potentially violent, other crimes against persons, sex, property, drugs, and minor offenses. For a defendant with multiple charges only the most serious offense was considered in data analyses.

Diagnoses

Diagnosis was coded using either the DSM-III-R or DSM-IV classification, depending on the year that the evaluation was completed. The diagnoses were divided into ten categories: psychotic disorder; organic disorder; mental retardation/learning disorder; affective disorder; anxiety/somatoform/conversion disorder; substance abuse disorder; personality disorder; paraphilia; and other. For those defendants with multiple diagnoses, the most severe disorder was used in data analyses. For example, if a defendant were diagnosed with both schizophrenia and a personality disorder, the former would be used in the data analyses.

Psychological Testing

Psychological tests were arranged into eight categories: personality/mood—objective; personality/mood—projective; personality/mood—projective and objective; cognitive; competence; malingering; sexual and other. Neuropsychological and neurological tests were entered as present or absent but were not further categorized.

Substance and Medication

Substance intake at the time of offense was categorized as alcohol, marijuana, cocaine/amphetamines, heroine/opiates, prescription medications, alcohol combined with another drug, combination of other drugs excluding alcohol, other/unknown, and none. Prescription medication was divided into seven categories: anti-psychotics, lithium, anti-depressants, anti-convulsants, anti-anxiety and combination of others, and others. If more than one category was prescribed, the most potent medication was coded. For example, if both Prozac and Haldol were prescribed to the defendant, Haldol, the anti-psychotic, would be used in data analyses.

Process Variables

Combining the time spent interviewing, collecting information, and report writing yielded the total time spent on the evaluation. All time variables were collapsed into categorical variables of either above or below the mean. We dichotomized these time variables both to facilitate interpretation and because our additional analyses examining these variables on a continuous level yielded similarly insignificant results (for a defense of dichotomization, see Farrington & Loeber, 2000).

Insanity Opinion

Opinion based on mental status at the time of the offense was derived using the three prongs of the insanity standard: (a) the ability to understand the nature, character, and consequences of the act; (b) the ability to distinguish right from wrong; and (c) the ability to resist the impulse of the act. This study does not include those evaluations in which the prongs—and hence the opinion—were not identified.

RESULTS

Table 1 summarizes information regarding the demographic and crime characteristics of the defendants opined to be either sane or insane by the evaluating

Table 1. Demographic and crime characteristics of defendant by opinion regarding sanity at the time of the offense

Variable	Sane	Insane	$\chi^2(1)$	CC
Age			21.24***	0.11
<36	852 (91%)	88 (9%)		
>36	609 (83%)	124 (17%)		
Gender			5.94*	0.04
Male	2867 (90%)	304 (10%)		
Female	568 (87%)	83 (13%)		
Minority status			6.88**	0.04
Yes	1489 (91%)	143 (9%)		
No	2008 (89%)	257 (11%)		
Type of crime				
Violent	1702 (89%)	208 (11%)	0.00	0.00
Potentially violent	717 (89%)	85 (11%)	0.01	0.00
Other crimes against persons	381 (86%)	61 (14%)	4.25*	0.03
Sex	365 (93%)	29 (7%)	5.45*	0.03
Property	733 (88%)	104 (12%)	2.46	0.02
Drugs	155 (96%)	8 (5%)	6.19*	0.03
Minor	195 (86%)	32 (14%)	2.54	0.02
Defendant's criminal history			7.76**	0.04
Prior convictions	2381 (90%)	256 (10%)		
No prior convictions	1568 (88%)	221 (12%)		

CC = contingency coefficient, a measure of effect size.
*$p<0.05$; **$p<0.01$; ***$p<0.0001$.
ns vary by analysis due to some missing data.

clinicians. As summarized in Table 1, significant differences were found on seven demographic and crime variables. However, even the largest effect size (CC = 0.11) is considered small (Cohen, 1992). Because 66% of the data was missing for age, it was not entered into subsequent analyses. Violent, potentially violent, property crimes, and minor crimes did not reflect significant differences and were not subsequently entered into the logistic regressions.

Table 2 summarizes the psychiatric characteristics of the defendants opined to be sane or insane by the evaluating clinician. The presence of a psychotic disorder emerged as the most relevant psychiatric characteristic, yielding an effect size (CC = 0.21) in the small to medium range (Cohen, 1992). Organic diagnoses, mental retardation/learning disorders, paraphilias, and anxiety/somatoform/conversion disorders did not reveal significant differences and were not subsequently entered into the logistic regression.

Table 3 summarizes certain process related characteristics of the evaluations conducted by psychologists versus psychiatrists. Although statistically significant, the effect sizes for the interdisciplinary differences reported in Table 3 tended to be small. *Post hoc* analysis of the differences in opinion by the two disciplines revealed that these differences actually reflected degree of experience ($\chi^2(1, N = 3471) = 8.90, p = 0.003$) by the various evaluators rather than actual disciplinary differences. Evaluators who had conducted less than the mean number ($n = 21$) of evaluations were more likely to opine that the defendant met the standard for insanity. In other words, less experienced evaluators were more likely to "find" legal insanity. Since a smaller number of experienced psychiatrists were present in this sample, the trend tended to show that psychiatrists were more likely to opine that a defendant met the standard for legal insanity.

Table 2. Psychiatric characteristics of the defendant at the time of the offense

Variable	Sane	Insane	$\chi^2(1)$	CC
Diagnosis received			21.75***	0.08
Yes	2635 (88%)	364 (12%)		
No	494 (95%)	27 (5%)		
Psychiatric diagnoses				
Psychotic disorders	823 (76%)	255 (24%)	229.22***	0.21
Organic disorders	106 (80%)	26 (20%)	10.86**	0.05
Mental retardation/learning disorders	314 (89%)	37 (11%)	0.04	0.00
Affective disorders	658 (87%)	99 (13%)	4.42*	0.03
Substance abuse	907 (97%)	26 (3%)	76.88***	0.12
Anxiety/somatoform/conversion	20 (100%)	—	2.45	0.02
Dissociative disorders	9 (69%)	4 (31%)	5.32*	0.03
Personality disorder	498 (96%)	20 (4%)	29.24***	0.08
Paraphilias	30 (100%)	—	3.68	0.03
Other	88 (97%)	3 (3%)	5.49*	0.03
Prior psychiatric history			72.36***	0.12
Prior hospitalization	2372 (87%)	398 (14%)		
No prior hospitalization	1732 (94%)	117 (6%)		
Prescribed medication at the time of the offense			83.12***	0.14
Yes	1324 (83%)	270 (17%)		
No	2617 (92%)	226 (8%)		
Taking medication at the time of the offense			6.80*	0.06
Yes	635 (88%)	85 (12%)		
No	1124 (84%)	215 (16%)		
Using substance at the time of the offense			64.68***	0.12
Yes	1601 (94%)	112 (7%)		
No	2164 (86%)	367 (15%)		

CC = contingency coefficient, a measure of effect size.
*$p < 0.05$; **$p < 0.001$; ***$p < 0.0001$.
ns vary by analysis due to some missing data.

Table 4 summarizes logistic regressions that were conducted to assess the cumulative value of the various significant variables in explaining differences in opinions regarding sanity and insanity. Three logistic regressions were conducted based upon the demographic, criminal, and psychiatric characteristics of the defendant. All significant variables and the experience variable were introduced into an overall model predicting an opinion supportive of insanity. The final outcome model was highly significant ($\chi^2(12, N = 1680) = 241.88, p < 0.000$) and contained nine variables. These include non-minority status; no drug offense charges; having a prior conviction; diagnosis of psychotic, organic, or affective disorder without a personality disorder diagnosis; previous psychiatric hospitalization; and not being under the influence of substances at the time of the offense. It is worth noting that the final model in Table 4 used the data subset that did not include the most prolific evaluators.

In an attempt to replicate findings from Cochrane et al. (2001), we ran a logistic regression including seven offense variables and ten diagnostic variables. Results were consistent with findings by Cochrane and colleagues in that there were no significant relationships between charges and psycholegal opinion once

Table 3. Process characteristics of sanity evaluations by clinical psychologists and psychiatrists

Variable	Clinical psychologist ($N=3121$)	Psychiatrist ($N=350$)	$\chi^2(1)$	CC
Opinion			11.84***	0.06
Insane	407 (13%)	69 (20%)		
Sane	2714 (87%)	281 (80%)		
Number of Evaluators			53.67***	0.12
Only one	2260 (74%)	182 (55%)		
More than one	801 (26%)	150 (45%)		
Hours spent				
Interviewing defendant	3.04 (2.18)	3.00 (3.36)	3.03	0.03
Collecting information	1.75 (1.89)	1.78 (2.23)	0.19	0.01
Report writing	3.43 (3.62)	3.39 (3.75)	0.07	0.00
Total time spent	8.28 (6.04)	8.21 (8.29)	2.55	0.03
Sources of information obtained				
Copy of warrant	2792 (91%)	288 (85%)	12.40***	0.06
Reasons for evaluation	2682 (87%)	313 (92%)	7.59**	0.05
Statements by the defendant	1207 (40%)	121 (37%)	0.97	0.02
Defendant's criminal history	1347 (45%)	156 (48%)	0.58	0.01
Information about alleged offense	2632 (86%)	264 (78%)	12.57***	0.06
Psychiatric/medical records	1906 (63%)	242 (72%)	10.52**	0.06
Witnesses statements	1322 (44%)	105 (32%)	15.23***	0.07
Use of psychological or neuropsychological tests			27.15***	0.12
Psychological	400 (22%)	11 (6%)		
Neuropsychological	185 (10%)	3 (2%)	15.03***	0.09
Neurological	21 (1%)	—	2.25	0.03

CC = contingency coefficient, a measure of effect size.
*$p<0.05$; **$p<0.01$; ***$p<0.0001$.
$n=3471$, a sample reduced from the overall dataset to control for the influence of a few prolific evaluators.

diagnoses were also considered. The model was highly significant ($\chi^2(17, N=5175)=332.23$, $p<0.0001$) and contained eight predictive variables. Unlike the previous study, however, our analysis did indicate one offense type—i.e. drug offense ($\beta=-0.42$ (0.20), $p=0.037$, OR = 0.43)—as being significantly, although negatively, related to an insanity opinion. Diagnoses significant in rendering an opinion of insanity included psychotic diagnoses ($\beta=0.68$ (0.07), $p<0.001$, OR = 3.86), organic diagnoses ($\beta=0.57$ (0.12), $p<0.001$, OR = 3.12), affective diagnoses ($\beta=0.31$ (0.08), $p\leqslant 0.000$, OR = 1.87), mental retardation/learning disorders diagnoses ($\beta=0.22$ (0.10), $p=0.031$, OR = 0.43), dissociative diagnoses ($\beta=0.82$ (0.31), $p=0.007$, OR = 5.18), and not having a substance abuse ($\beta=-0.52$ (0.11), $p<0.001$, OR = 0.56) or a personality disorder diagnosis ($\beta=-0.35$ (0.13), $p=0.005$, OR = 0.49).

An additional question involved whether the proportion of defendants opined to be insane remained consistent over the 10 year period during which these evaluations were conducted. As there were no changes during this period in the laws governing the sanity evaluations or the sanity standard itself, the degree of consistency was explored as a crude measure of the temporal stability of the opinions offered to the courts by this large cohort of primarily community based evaluators. An ANOVA of the percentages of defendants opined to be insane over each year was non-significant ($\chi^2(1, N=3638)=2.44$, $p=0.118$) suggesting no meaningful

Table 4. Logistic regression summary for demographic, offense characteristics, psychiatric diagnoses and overall model for predicting an insanity opinion

Variable	β	SE	p	OR
Demographic variables ($N=3524$)				
Female	0.15	0.07	0.026	1.36
Minority	−0.13	0.06	0.032	0.78
Offense variables ($N=4426$)				
Drug offense	−0.47	0.20	0.016	0.39
Prior conviction	−0.14	0.05	0.004	0.75
Psychiatric variables ($N=2813$)				
Diagnosis received	0.31	0.14	0.032	1.86
Psychotic diagnosis	0.51	0.11	0.000	2.76
Organic diagnosis	0.46	0.18	0.009	2.52
Affective diagnosis	0.24	0.12	0.041	1.62
Substance abuse diagnosis	−0.47	0.16	0.003	0.39
Personality disorder	−0.45	0.17	0.007	0.40
Prior hospitalization	0.27	0.08	0.001	1.73
Under the influence at the time of the offense	−0.33	0.07	0.000	0.52
Overall model* ($N=1680$)				
Minority	−0.23	0.09	0.010	0.63
Drug offense	−1.13	0.51	0.029	0.11
Prior conviction	−0.20	0.08	0.016	0.67
Psychotic diagnosis	0.87	0.14	0.000	5.66
Organic diagnosis	0.48	0.22	0.029	2.61
Affective diagnosis	0.43	0.15	0.004	2.39
Personality disorder	−0.47	0.22	0.031	0.39
Prior hospitalization	0.32	0.10	0.001	1.90
Under the influence at the time of the offense	−0.38	0.09	0.000	0.47

*This model includes all the previous variables and the variable representing evaluator experience.

changes in the frequency of and insanity opinion offered the courts throughout the 10 years of data collection and the large number of evaluations/evaluators.

Finally, Table 5 summarizes the various prongs of the insanity defense coded by the evaluators as supporting their psycholegal opinion of insanity. As summarized, of the 563 individuals opined to be insane by the evaluators, 91% of these involved an opinion that encompassed the two cognitive prongs of the insanity standard either singularly, in combination, or in conjunction with recognition of the defendant's inability to resist the impulse to act. In only 51(9%) instances did the evaluating clinician identify irresistible impulse as the only prong of the insanity defense relevant to the legal question.

DISCUSSION

These findings are highly consistent in many respects with clinical wisdom concerning the evaluation of sanity at the time of the offense. The overall model that best predicted an opinion supportive of insanity included various serious Axis I diagnoses, the absence of an Axis II disorder as the primary diagnosis, prior psychiatric hospitalizations, not being under the influence of substances at the time of the offense, and not being charged with a drug offense. Clearly, and not surprisingly, a history of serious mental illness was the most influential clinical

Table 5. Prongs cited in evaluations supportive of insanity

Type of impairment found	
Impairment on all three prongs, A, B, C* (both cognitive and volitional)	223 (40%)
Impairment on both A and B (solely cognitive)	119 (21%)
Impairment on both A and C (both cognitive and volitional)	22 (4%)
Impairment on both B and C (both cognitive and volitional)	20 (4%)
Impairment on A only (solely cognitive)	58 (10%)
Impairment on B only (solely cognitive)	70 (12%)
Impairment on C only (solely volitional)	51 (9%)

*The three prongs of the insanity standard: (A) the ability to understand the nature, character, and consequences of the act; (B) the ability to distinguish right from wrong; and (C) the ability to resist the impulse of the act.

factor, with individuals with a psychotic diagnosis being over five times more likely to be found insane than those without this type of diagnosis. As in the Cohrane et al. (2001) study, clinical characteristics of the defendant were found to override offense characteristics in predicting an opinion of insanity, except—in our sample—for a negative association with drug offenses. The professional discipline of the evaluators did not remain significant in this final model, suggesting that well trained forensic psychiatrists and psychologists, while using a slightly different methodology, are fairly consistent in terms of the proportion of defendants they opine to meet the insanity standard.

Evaluators often offered their opinions on the basis of incomplete data. As shown in Table 3, in more than half the cases, evaluators of both professional disciplines were offering sanity opinions without having seen statements by the defendant, the defendant's criminal history, and/or the statements of witnesses. In a smaller proportion of cases, evaluators had not obtained information about the alleged offense. There are many cases in which a screening evaluation can be completed without these data, and some hospital-based evaluators may feel that the inpatient setting affords enough first-hand data to make collateral data less necessary. However, the forensic evaluator who reaches an opinion about criminal responsibility without looking at these sources of information is at greater risk of reaching the wrong conclusion and would have great difficulty defending this omission on vigorous cross-examination. He or she may appear even more negligent given the practice standards implied by survey research, which indicates that most forensic professionals view such collateral information as necessary for conducting a sanity evaluation (Borum & Grisso, 1996). More broadly, when many forensic evaluators fail to consider relevant data, they leave the field vulnerable to the types of criticism that Grisso (1986) summarized nearly two decades ago: specifically, that evaluators fail to consider clinically and legally relevant information. In fairness to these examiners, it should be noted that we are aware of some jurisdictions within the state wherein prosecutors consistently decline requests to share non-statutorily-mandated offense-related information with forensic evaluators, presumably out of concern that this information will be shared with defense counsel. Although such practice does not violate the rules of discovery, it may make it more difficult for evaluators to reach an accurate psycholegal opinion in certain cases.

One finding of broad societal concern is that minority status had a statistically significant negative association with an opinion of insanity, even when all the relevant variables were controlled within the final outcome model. Although the effect size was small (CC = 0.04, reflecting 8.5% of minorities versus 11.4% of whites opined to be insane), the discrepancy is important in light of the longstanding racial disparity within the American criminal justice system. Racial bias was not observed in the percentage of defendants referred for evaluation: 43% minority and 57% white defendants were referred for evaluation as compared to 42% minority and 59% white suspects arrested in Virginia (Virginia Department of State Police, 2001), suggesting that racial dynamics are influential in the outcome of the forensic evaluation rather than in the process of representation and referral by counsel. It was not possible to determine whether the racial disparity in our sample occurred only across evaluator/defendant racial lines, as data were not available on the race of the referring attorney or evaluator, although it is clear that the majority of evaluators were of non-minority status.

All models reflected a negative relationship between the use of substances at the time of the offense and a psycholegal opinion supportive of insanity. While this finding likely reflects the legal responsibility that derives from voluntary intoxication, it may also reflect the connotation of wrong-doing and, therefore, culpability in instances wherein substances are paired with criminal behavior. Apparently, there are cases in which evaluators of psychiatrically impaired individuals who offended while under the influence of a substance have the opinion that the influence of the substance overrode the psychotic thinking that might otherwise have been relevant to the criminal behavior, whether or not it is the pivotal factor in the motivation for the crime. The methods used here do not permit an analysis of the possible explanations for this finding, which could include a reasoned opinion that the underlying illness did not cause the defendant to meet the legal test of insanity, a moral bias against substance abusers, and/or the biasing effect of anticipating that juries are unlikely to negate culpability in instances in which substance abuse is involved.

The analyses that were conducted across the 10 years of data collection suggest no significant changes in the proportion of defendants opined to be insane from one year to the next. These numbers are noteworthy given the lengthy time period and the numerous evaluators, which make for a sample size that should be sensitive even to modest year-to-year changes in opinion rates. These results suggest that a community-based forensic mental health system is able to offer the courts a reliable cohort of forensic evaluators. The consistent base rate of insanity opinions across years in Virginia, which varied from 7 to 15% with a mean of 12%, is also consistent with national trends indicating that findings of insanity are unusual regardless of locale (Warren, Rosenfeld, Fitch, & Hawk, 1997). It is likely that the consistency observed within the Commonwealth of Virginia derives in part from the rigorous nature of the training required of forensic evaluators prior to their involvement in the community-based system. This multi-disciplinary training, which is mandated by statute and funded by the state Department of Mental Health, is within a university environment, integrates legal knowledge with forensic principles, and does so in an applied manner that is designed to build upon the preexisting clinical experiences of the trainees.

The analysis of the insanity "prongs" upon which the forensic evaluators based their findings of insanity documents the relative importance of the cognitive, as contrasted with the volitional, test of insanity. In 91% of the opinions supporting insanity that were formulated by this cadre of evaluators, examiners identified at least one of the two cognitive prongs of the insanity standard as relevant to the opinions they offered to the courts. This pattern appears to reflect the longstanding difficulty that clinicians have in differentiating between an irresistible impulse and an impulse not resisted. Psychotic delusions are often central to a determination that the defendant was not able to differentiate right from wrong at the time of an offense and reflect a more clear-cut set of symptoms than are available to address the behavioral inhibitions implied by the volitional test. Bonnie suggested in 1983 that "whatever the precise terms of the volitional test, the question is unanswerable, or it can be answered only by 'moral guesses'" (p. 196).

From a different perspective, however, one could argue that, in the majority (57%) of cases, the evaluating clinician identified the volitional prong as relevant to their opinion of legal insanity. It is our impression that, in those cases in which the volitional prong was combined with the cognitive prong (48% of all cases), the evaluators perceived the cognitive prong to be the determinative factor for reaching an opinion supportive of insanity, with the volitional prong reflecting more the emotional nature of the crime versus an actual compelling state of volitional dyscontrol. The current data set, however, does not allow us to examine this hypothesis or to otherwise compare the two different interpretations of this data.

The findings reported here may be more representative than previous studies because they reflect the work of over 200 forensic examiners. Attempts were made to control for the biases of a few prolific examiners, but the lack of standardization and reliability in the actual evaluations are one substantial limitation of our data. Nevertheless, these findings speak to questions of public policy, by supporting the viability and reliability of community-based forensic training, and to questions of forensic practice, reflecting the factors that forensic psychologists and psychiatrists consider when offering opinions to the court.

REFERENCES

Bonnie, R. (1983). The moral basis of the insanity defense. *American Bar Association Journal, 69*, 194–204.
Borum, R., & Grisso, T. (1995). Psychological test use in criminal forensic evaluations. *Professional Psychology: Research and Practice, 26*, 465–473.
Borum, R., & Grisso, T. (1996). Establishing standards for criminal forensic reports: An empirical analysis. *Bulletin of the American Academy of Psychiatry and the Law, 24*, 297–317.
Callahan, L. A., Steadman, H. J., McGreevy, M. A., & Clark-Robins, P. (1991). The volume and characteristics of insanity defense pleas: An eight-state study. *Bulletin of the American Academy of Psychiatry and Law, 19*, 331–338.
Cochrane, R. E., Grisso, T., & Frederick, R. I. (2001). The relationship between criminal charges, diagnoses, and psycholegal opinions among federal pretrial defendants. *Behavioral Sciences and the Law, 19*, 565–582.
Cohen, J. (1992). A power primer. *Psychological Bulletin, 112*, 155–159.
Cooper, D. K., & Grisso, T. (1997). Five year research update (1991–1995): Evaluations for competence to stand trial. *Behavioral Sciences and the Law, 15*, 347–364.
Daubert, V. Merrell Dow Pharmaceuticals, 509 U.S. 579 (1993).
Farrington, D. P., & Loeber, R. (2000). Some benefits of dichotomization in psychiatric and criminological research. *Criminal Behaviour and Mental Health, 10*, 100–122.
Grisso, T. (1986). *Evaluating competencies*. New York: Plenum.
Grisso, T. (1992). Five year research update (1986–1990): Evaluations for competence to stand trial. *Behavioral Sciences and the Law, 10*, 353–369.
Hart, S. D., & Hare, R. D. (1992). Predicting fitness to stand trial: The relative power of demographic, criminal, and clinical variables. *Forensic Reports, 5*, 53–65.

Heilbrun, K. (1990). Response style, situation, third-party information and competency to stand trial: Research issues in practice. *Law and Human Behavior, 14*, 193–196.

Heilbrun, K. (1992). The role of psychological testing in forensic assessment. *Law and Human Behavior, 16*, 257–272.

Heilbrun, K., & Collins, S. (1995). Evaluations of trial competency and mental state at time of offense: Report characteristics. *Professional Psychology: Research and Practice, 26*, 61–67.

Heilbrun, K., Rosenfeld, B., Warren, J., & Collins, S. (1994). The use of third party information in forensic assessments: A two state comparison. *Bulletin of the American Academy of Psychiatry and Law, 22*, 399–406.

Heilbrun, K., Warren, J., & Picarello, K. (2003). The use of third party information in forensic assessment. In. I. B. Weiner, & A. M. Goldstein (Eds.), *Comprehensive handbook of psychology*, Vol. 11, *Forensic Psychology* (pp. 69–86). New York: John Wiley & Sons Ltd.

Kumho Tire Company v. Carmichael, No. 97-1079, 1999 U.S.LEXIS 2189 (March 23, 1999).

Melton, G., Petrilla, J., Poythress, N., & Slobogin, C. (1997). *Psychological evaluations for the courts* (2nd ed.). New York: Guilford.

Nicholson, R. A., & Kugler, K. E. (1991). Competent and incompetent criminal defendants: A quantitative review of comparative research. *Psychological Bulletin, 3*, 355–370.

Nicholson, R. A., & Norwood, S. (2000). The quality of forensic psychological assessments, reports, and testimony: Acknowledging the gap between promise and practice. *Law and Human Behavior, 24*, 9–44.

Otto, R. K., & Heilbrun, K. (2002). The practice of forensic psychology: A look toward the future in light of the past. *American Psychologist, 57*, 5–18.

Quinnell, F. A., & Bow, J. N. (2001). Psychological tests used in child custody evaluations. *Behavioral Sciences and the Law, 19*, 491–501.

Petrella, R. C., & Poythress, N. G. (1983). The quality of forensic evaluations: An interdisciplinary study. *Journal of Consulting and Clinical Psychology, 51*, 76–85.

Redding, R. E., Floyd, M. Y., & Hawk, G. L. (2001). What judges and lawyers think about the testimony of mental health experts: A survey of the courts and the bar. *Behavioral Sciences and the Law, 19*, 583–594.

Rogers, R., & Shuman, D. W. (2000). *Conducting insanity evaluations*. New York: Guilford.

Rosenfeld, B., & Ritchie, K. (1998). Competence to stand trial: Clinician reliability and the role of offense severity. *Journal of Forensic Science, 43*, 151–157.

Rosenfeld, B., & Wall, A. (1998). Psychopathology and competence to stand trial. *Criminal Justice and Behavior, 25*, 443–462.

Skeem, J. L., Golding, S. L., Cohn, N. B., & Berge, G. (1998). Logic and reliability of evaluations of competence to stand trial. *Law and Human Behavior, 22*, 519–547.

Slovic, P., Monahan, J., & MacGregor, D. G. (2000). Violence risk assessment and risk communication: The effects of using actual cases, providing instruction, and employing probability versus frequency formats. *Law and Human Behavior, 24*(4), 271–296.

U.S. Department of Justice. (1997). *Correctional population in the United States*. Washington, DC: Bureau of Justice Statistics.

U.S. Department of Justice. (2000). *Crime in the United States: Uniform crime reports*. Washington, DC: Federal Bureau of Investigation.

Virginia Department of State Police. (2001). *Crime in Virginia: Virginia uniform crime reporting program*. Richmond, VA: Author.

Warren, J. I. (1992). *Forensic evaluation reimbursement study: The Virginia Supreme Court*. Confidential report compiled for the Department of Mental Health, Mental Retardation and Substance Abuse.

Warren, J. I., Fitch, W. L., Dietz, P. E., & Rosenfeld, B. D. (1991). Criminal offense, psychiatric diagnoses, and psycholegal opinion: An analysis of 894 pretrial referrals. *Bulletin of the American Academy of Psychiatry and the Law, 19*, 63–69.

Warren, J. I., Rosenfeld, B., Fitch, W. L., & Hawk, G. (1997). Forensic mental health clinical evaluation: An analysis of interstate and intersystemic differences. *Law and Human Behavior, 21*, 377–390.

"Imagination, personality, and imaginary companions," Gleeson, Jarudi, Cheek, *Social Behavior and Personality*, 31 (7):721–738, 2003.[1]

1. Describe the subjects in this research. How does this relate to the author's affiliation? What problems does this pose for the conclusions of the study?
2. Under Imagination on p. 726, the paper says, "The alpha coefficient ... was .82." This refers to Cronbach's alpha, which measures the consistency of questions on a questionnaire.
3. The authors mention Table 1 in only two sentences starting at the bottom of p. 728. They say that the correlations indicate that there were several different aspects of imagination being assessed. What is it about Table 1 that leads them to this conclusion? What would Table 1 look like if their conclusion were not true?
4. At the bottom of Table 1, it says "for $r>.20$, $p<.05$." How did they determine this? Is there any issue presented by the fact that Table 1 contains 15 different correlations? Does this change their conclusions about Table 1?
5. Table 2 has a column labeled "d," which is said to be the "effect size" on p. 728. This is Cohen's d, which is how many standard deviations apart the averages are. But the standard deviations for the two groups are different. Which was used to compute d? Does this suggest what kind of t test was done, equal variance or unequal variance?
6. How does the effect size help you interpret Table 2? About what percent of those without an imaginary companion score higher in Imagery than the average subject who did have an imaginary companion?
7. In Table 2, the means for Hostile Daydreaming and Vivid Dreams are rather similar, but the P-value for Hostile Dreaming is twice that for Vivid Dreams. Why is this?
8. Table 3 lists correlations. Why?
9. There are only two significant differences in Table 4. What do the authors have to say about this?

---

[1] © *2003, Social Behavior and Personality: An International Journal. Used by permission.*

IMAGINATION, PERSONALITY, AND IMAGINARY COMPANIONS

TRACY R. GLEASON, RACEEL N. JARUDI, AND JONATHAN M. CHEEK
Wellesley College, Wellesley, MA, USA

A sample of 102 college women completed a set of imagination and personality measures and reported whether they had ever had imaginary companions during childhood. Participants who reported imaginary companions scored higher than did those who did not on measures of imagination including imagery use, hostile daydreams, and vivid night dreams, and on personality scales including dependent interpersonal styles and internal-state awareness. Participant groups did not differ significantly on shyness, other interpersonal styles, or measures of self-concept. Comparison of these results with research on children and adolescents with imaginary companions suggests a coherent developmental pattern in social orientation characterized by sensitivity and accommodation to others' needs.

A common question addressed in the literature on imaginary companions concerns whether individuals who create pretend friends differ from those who do not. Some investigations (e.g., Manosevitz, Prentice, & Wilson, 1973) have examined features of children's social environments, hypothesizing that having fewer available playmates, no siblings, or having lost a parent might prompt a child to create an imaginary friend. These investigations have not consistently distinguished between those with and without imaginary companions except with respect to birth order: compared to children without imaginary companions, children with imaginary companions are more frequently only and first-born children or those with siblings far apart in age (Bouldin & Pratt, 1999; Singer &

Tracy R. Gleason, Raceel M. Jarudi, and Jonathan M. Cheek, Department of Psychology, Wellesley College, Wellesley, MA, USA.
This study was supported by a Wellesley College Faculty Award to Jonathan M. Cheek. The authors thank Amy Gower for her comments on an earlier draft of the manuscript.
Appreciation is due to reviewers including: Dr. David Pearson, Child and Adolescent Mental Health Services, The Health Clinic, 7 Pyle Street, Isle of Wight, PO30 1JW, UK <david.pearson@iow.nhs.uk> and Paula Bouldin, PhD, School of Psychology, Deakin University, Geelong, Victoria, Australia 3217 <paulab@deakin.edu.au>
Keywords: imaginary companion, imagery, shyness, personality, interpersonal style.
Please address correspondence and reprint requests to: Tracy R. Gleason, Department of Psychology, Wellesley College, Wellesley, Massachusetts 02481-8203 USA. Phone: (781) 283-2487; Fax: (781) 283-3730; Email:<tgleason@wellesley.edu>

Singer, 1990). Nevertheless, even this finding is not ubiquitous (Hurlock & Burstein, 1932). Consequently, some researchers have addressed the role of intrapsychic features of the individual in the development of imaginary companions. Two primary factors have been proposed: imagination and personality.

IMAGINATION

Many creators of imaginary companions report that their pretend friends are accompanied by vivid auditory and visual imagery (Dierker, Davis & Sanders, 1995; Hurlock & Burstein, 1932). Empirical evidence suggests that the presence of imagery of this kind might be indicative of individual differences in imagery use and type. For example, Meyer and Tuber (1989) found that the Rorschach scores of 4- and 5-year-old children with imaginary companions are indicative of superior symbolic representation and unusual imaginative resources compared to norms established for this age group. Although the capacity to engage in imagined dialogues with others is universal, and most individuals – even those without pretend friends – spend time engaged in this type of fantasy (Caughey, 1984; Watkins, 2000), those who create imaginary companions might be particularly prone to engaging in fantasy. Evidence for such group differences has emerged in research with both children and adults.

In infancy, children who later develop imaginary companions show a preference for fantasy- over reality-based toys (Acredolo, Goodwyn, & Fulmer, 1995). These differences have also emerged between preschool-aged children with and without imaginary companions. According to parent reports, young children with such friends are more likely to incorporate myth into their play, to explain events as magical, to play imaginatively (Bouldin & Pratt, 1999) and to engage in pretense in an experimental situation than are their peers (Taylor, 1999).

Children with and without imaginary companions may also differ in concentration and focusing attention, skills that might tap imaginative abilities. For example, Singer (1961) evaluated 6- to 9-year-old high- and low-fantasy children's ability to wait patiently during a game. Children classified as high-fantasy, a category partly determined by whether or not a child had an imaginary companion, were able to wait longer than were those classified as low-fantasy. Singer proposed that the patience of the high-fantasy children might be related to their use of imagination to distract themselves while waiting. Mauro (1991) also found differences in this kind of attentional focusing between children with and without pretend friends in a sample of 4-year-olds, but in a follow-up session when the children were 7 these differences had disappeared. Differences in attentional focusing were also not replicated in another study of children (average age 5.9) with and without pretend friends (Manosevitz, Fling, & Prentice, 1977).

While findings for differences in imaginative skills in children with and without imaginary companions might be ambiguous, research on adolescents and adults suggests that any variation present in childhood becomes more pronounced with development. For example, some adolescents create imaginary companions to whom journal entries are addressed, and the best predictor of this behavior is a propensity for daydreaming (Seiffge-Krenke, 1993; 1997). In two self-report studies involving comparisons between college students with and without memories of imaginary companions, Dierker, Davis and Sanders (1995) found differences on measures of imaginative involvement, such as interest in and use of fantasy and belief in the paranormal, among women and on measures of normative dissociation among both genders. Differences also emerged between women with varying degrees of vividness in their memories of their imaginary companions. High-vividness participants, who reported seeing and hearing their companions and believing they were real, had significantly higher imaginative involvement than did either low vividness participants or women who had not had imaginary companions as children.

In summary, research on children with and without imaginary companions suggests that children with imaginary companions may be predisposed to engaging in fantasy play and may have heightened imaginative abilities in comparison to their peers. The strength of this difference appears to be moderate in childhood but to either increase with development or be stronger in adults, especially women, who remember their childhood companions. Consequently, we hypothesized that women who had recollections of childhood imaginary companions would score higher on measures of vividness of imagery and dreams. We also hypothesized that they would engage in more daydreaming than did adults without recollections of pretend friends. Given that it has not been studied previously, we were interested to see also if the content of day and night dreams differed between participant groups. Lastly, we investigated whether the imaginary companion group was more prone than were their peers to become involved in externally created fantasies, such as those provided by reading fiction or watching movies.

Personality

In addition to differences in imagination, personality characteristics have been examined to explain why only some individuals create imaginary companions. The creation of a pretend friend has been described as an aid for children who are particularly shy or neurotic (Nagera, 1969) or who enjoy having someone to boss around (Vostrovsky, 1895). Many early studies of children with imaginary companions describe them as having personality problems, such as timidity or bossiness (Svendsen, 1934), a "nervous temperament" (Vostrovsky, p. 396), or an inability to play well with other children as a result of temperament (Ames & Learned, 1946). However, as noted by Taylor (1999), these studies are often flawed. In some cases children with imaginary companions are not compared to children without (Ames & Learned, Vostrovsky), or are compared to unmatched control groups (Svendsen). In other cases, children with imaginary companions are recruited from clinical settings to which they have been referred for some sort of problem (e.g., Ames & Learned).

More recent work on the personalities of children with imaginary companions avoids the methodological pitfalls of earlier work and paints a particularly social view of the personalities of imaginary companion creators. For example, Partington and Grant (1984) noted higher scores on an index of social interaction for children with imaginary companions versus those without. Their explanation of this finding is that children with imaginary companions have an orientation that is toward rather than away from contact with others. Support for this interpretation comes from Bouldin and Pratt (2002), who compared children on several measures of anxiety, temperament, and specific fears. They found that children with imaginary companions experienced more concentration anxiety than did their peers, meaning that they had concerns regarding meeting others' expectations and may have been particularly attuned to interactions with others. Of course, not every child with an imaginary companion is socially oriented, but even so, low-social children with imaginary companions demonstrate more complex pretend play and obtain higher scores on measures of solutions to social problems in comparison to low-social children without imaginary companions (Bach, Chang, & Berk, 2001).

Further evidence of personality differences suggests that children with imaginary companions are less fearful and less generally anxious (Singer & Singer, 1990), more cooperative (Partington & Grant, 1984; Singer & Singer), and better at talking and initiating activities with adults (Manosevitz et al., 1973) than are their peers. Mauro (1991) noted no personality differences on a temperament measure between 4-year-old children with and without imaginary companions except that children with such friends appeared to be better at focusing attention, as mentioned above, and less shy. However, again, these differences had disappeared by the time children were 7, and other researchers have also found no differences in temperament between children with and without imaginary companions (Bouldin & Pratt, 2002; Manosevitz et al., 1973). Taken together, this research suggests that in terms of personality, children with and without imaginary companions seem more similar than different (Taylor, 1999), although children with pretend friends may be more sociable than are those without, particularly with adults.

Research on adolescent creators of diary-based imaginary companions supports the notion that individuals who create such friends may be socially oriented. Aside from readiness for daydreaming, the best indicator of the formation of a diary companion was an active coping style, in which one's social resources are actively sought for help and advice. The creation of a diary imaginary companion was also predicted, albeit less strongly, by a positive self-concept and high self-esteem in particular (Seiffge-Krenke, 1993; 1997).

Retrospective reports also contribute to the examination of the relation between personality and the creation of an imaginary companion. Adolescents who created imaginary companions in childhood had higher levels of psychological distress than their peers did. These teens had difficulty handling emotions and were more anxious about social interaction, although they did not differ from their peers on a general measure of well-being (Bonne, Canettie, Bachar, DeNour, & Shalev, 1999). While these findings appear to run contrary to those of the diary studies, if individuals who remember imaginary companions are particularly socially oriented, these results may reflect the unusually strong impact of the social stresses characteristic of adolescence on these socially attentive adolescents. Support for this interpretation comes from Wingfield's (1948) comparison of women who had memories of imaginary companions with a normative sample on a set of personality measures. Similarly to the adolescents with diary companions, women who had created imaginary companions sought advice and encouragement from others more than average, and they were rated as disliking solitude and being less introverted and neurotic than was the average college woman.

Overall, the literature suggests that personality differences between those with and without imaginary companions might be expected in measures assessing an individual's social orientation. Consequently, we chose several personality measures related to relationships. First, we hypothesized that compared to those without, individuals with memories of imaginary companions would have interpersonal styles that were similar to the orientation that Horney (1945) called "moving toward others." Second, because children with imaginary companions seem better able to focus their attention than are other children, we predicted higher scores for those reporting imaginary companions on measures of self-focused attention (Fenigstein, Scheier, & Buss, 1975). Third, one personality variable related to relationships with others that has not been studied with respect to imaginary companions is the impostor phenomenon (Clance, 1985). We speculated that the heightened social orientation of those with imaginary companions, in combination with an imaginative inner focus, might make them particularly susceptible to a sense of disconnection between their public and private selves; thus they may be more likely to feel like impostors in their social interactions, and perhaps even be prone to the self-preoccupation of hypersensitive narcissism (Hendin & Cheek, 1997).

Although shyness appears uncharacteristic of children with imaginary companions (Taylor, 1999), the theoretical assumption that shy children and adults have rich inner lives and "hidden talents" is quite prevalent (Avila, 2002; Hall, 1982). Because shyness and imaginary companions have not been investigated in adults, we included a measure of shyness but hypothesized that those who could and could not remember pretend friends would not differ. Finally, a measure of positive childhood memories was included as a proxy for personality development as well as a check on the possibility that the general childhood memories of individuals with imaginary companions are distinctly different from those of individuals without imaginary companions.

METHOD

PARTICIPANTS

Participants were 102 female college students aged 18-21, who participated as part of course credit for introductory and intermediate level psychology classes and completed a questionnaire packet that included the measures described below. We recruited women because previous retrospective reports have been focused on female participants (Dierker et al., 1995; Wingfield, 1948) and because more women than men recall having an imaginary companion as a child. Approximately 70% of participants identified their ethnicity as Caucasian or European-American, and there were too few participants from any other ethnic group to analyze the data for ethnic differences. Participants completed the questionnaire packet individually in the Psychology Department. Afterwards, they were provided with an explanation of the research.

MEASURES

Our measures of imagination related to imagery use, daydreams, and night dreams, and our personality measures assessed shyness, social orientation, the impostor phenomenon, narcissism, self-consciousness, and happy childhood memories. All items were evaluated on a 5-pt. scale (1 = *strongly disagree* to 5 = *strongly agree*) unless otherwise specified, and measures of internal consistency are reported using alpha coefficients.

Imagination Imagery use was assessed using 10 items adapted from the Habitual Use of Imagery and the Imagination factors of Paivio's Individual Differences Questionnaire (Paivio & Harshman, 1983). The Habitual Use factor included items such as "I often use mental images or pictures to help me remember things," and the Imagination factor included items such as "My powers of imagination are higher than average." The alpha coefficient for the two factors together was .82 (cf. Hiscock, 1978).

Daydreaming frequency (e.g., "Whenever I have time on my hands, I daydream") was measured using a 12-item subscale of the Imaginal Processes Inventory (Singer & Antrobus, 1970). Content of daydreams was assessed via the scales and subscales of the Short Imaginal Processes Inventory: Positive-Constructive Daydreaming and Guilty-Dysphoric Daydreaming (Huba, Singer,

Aneshensel, & Antrobus, 1982). Hostile daydreaming (e.g., "I find myself imagining ways of getting even with those I dislike") was assessed with a 3-item index derived from the Guilty-Dysphoric Daydreaming factor (Singer & Bonanno, 1990). The alpha coefficients for these daydreaming scales were .92, .86, .84, and .80, respectively.

Concerning night dreams, two items were included: "When I dream, I dream in color" and "My night dreams are extremely vivid." These items were correlated $r = .48$ ($p < .01$) and were thus combined to form an index of vivid and colorful night dreams.

The last imagination variable, fantasizing about fictional characters (e.g., "I really get involved with the feelings of the characters in a novel"), was assessed with the 7-item Fantasy Scale of the Interpersonal Reactivity Index (alpha = .80) (Davis, 1983).

Personality Shyness was assessed using a 20-item version of the Cheek and Buss (1981) Shyness Scale, which measures somatic (e.g., "I feel tense when I'm with people I don't know well"), cognitive (e.g., "I feel painfully self-conscious when I'm around strangers") and behavioral (e.g., "I am socially somewhat awkward") components of shyness (Cheek & Krasnoperova, 1999; Cheek & Melchior, 1985). The alpha coefficient for this scale was .94.

Karen Horney (1945) identified the three core interpersonal styles of moving toward, away from, and against other people (for a recent review, see Waters & Cheek, in press). We measured Horney's styles using the personality scales selected by Cheek and Krasnoperova (1999). Moving toward others was measured using the 17 items of two subscales of the Robins et al. (1994) Personal Style Inventory (PSI)-Dependency (e.g., "I find it difficult to be separated from people I love") and Pleasing Others (e.g., "I often put other people's needs before my own"). The alpha coefficient for this measure was .85. Moving away from others (e.g., "I tend to keep other people at a distance") was assessed with Defensive Separation, a 12-item subscale of the PSI (alpha = .79). The aggressive interpersonal style of moving against others was assessed using 14 items derived from two subscales of the Buss and Perry (1992) Aggression Questionnaire-Physical Aggression (e.g., "Once in a while, I can't control the urge to strike another person") and Verbal Aggression (e.g., "I tell my friends openly when I disagree with them"). The alpha coefficient for the resulting index was .82.

The impostor phenomenon (e.g., "I have often succeeded on a test or task even though I was afraid that I would not do well before I undertook the task") was assessed with Clance's (1985) Impostor Phenomenon Scale, a 20-item measure of self-doubt and feelings of inauthenticity (alpha = .92).

We included the Hypersensitive Narcissism scale (Hendin & Cheek, 1997), a 10-item measure of anxious self-preoccupation (e.g., "I can become entirely absorbed in thinking about my personal affairs, my health, my cares or my relations to others"). The alpha coefficient was .76.

The Public Self-Consciousness scale of the Fenigstein, Scheier, and Buss (1975) Self-Consciousness Inventory, including two items revised by Scheier and Carver (1985) was used to assess habitual attentiveness to social aspects of the self (e.g., "I am usually aware of my appearance"). The scale had an alpha coefficient of .81.

The original Private Self-Consciousness scale contains items that refer both to self-reflection and to self-perception, and these items suggest two factors of subjective experience (Burnkrant & Page, 1984; Piliavin & Charng, 1988). Therefore, we included two scales to assess both introspectiveness and internal state awareness. Introspectiveness (e.g., "I think about myself a lot") included the seven self-reflection items from the Private Self-Consciousnes scale, as revised by Scheier and Carver (1985), and nine items added from Guilford's (1959) measurement of thinking introversion by Cheek and Madaffari (1998). The resulting 16-item measure produced an alpha coefficient of .88. The second factor of Private Self-Consciousness, habitual attentiveness to internal feelings and emotions (e.g., "I generally pay attention to my inner feelings"), was assessed with two items from the Private Self-Consciousness scale (Scheier & Carver, 1985) one item from the Private Body Consciousness scale (Miller, Murphy & Buss, 1981), and one item that was added by Cheek and Madaffari. This 4-item Internal State Awareness scale had an alpha coefficient of .71.

Happy childhood memories (e.g., "Overall, I would say that I had a happy childhood") were assessed using a 20-item version of the Assessing the Wounds of Childhood checklist (Aron, 1996; Aron & Aron, 1997). The resulting measure produced an alpha coefficient of .90.

Imaginary companion On the final page of the questionnaire packet, memory of a childhood imaginary companion was assessed using one question drawn from Schaefer (1969): "As a child, did you ever have any imaginary companions (e.g., friends, animals)?"

RESULTS

Of the 102 participants in the study, 29.4% (n = 30) reported recollections of childhood imaginary companions. A series of independent t-tests compared the scores of this group against those of the remainder of the sample on the imagination and personality variables, and the effect size d was computed for each t-test (Cohen, 1988).

IMAGINATION

The correlations among the seven measures of imagination are presented in

Table 1. The correlation between measures ranged from -.14 to .74 and averaged .27, which indicates that several somewhat distinct aspects of imagination were assessed. The mean scores for participants who did and did not report childhood imaginary companions appear in Table 2. As expected, the scores of the imaginary-companion group were higher than were those of their counterparts for every variable, and these differences reached statistical significance in several

TABLE 1
CORRELATIONS AMONG MEASURES OF IMAGINATION

	Imagery use	Daydreaming frequency	Positive daydreams	Guilty daydreams	Hostile daydreams	Vivid colorful night dreams
Daydreaming frequency	.26					
Positive-constructive daydreams	.28	.33				
Guilty-dysphoric daydreams	.21	.21	-.14			
Hostile daydreams	.28	.26	.01	.74		
Vivid colorful night dreams	.65	.14	.28	.05	.19	
Fantasy scale	.48	.41	.33	.13	.14	.33

$N = 102$, for $r > .20$, $p < .05$

TABLE 2
GROUP DIFFERENCES ON MEASURES OF IMAGINATION

	Imaginary companion $n = 30$		No imaginary companion $n = 72$				
	M	SD	M	SD	t	d	p
Imagery use	42.30	6.65	39.29	5.63	2.33	.50	.02
Daydreaming frequency	41.63	8.41	38.96	8.60	1.44	.31	.15
Positive-constructive daydreaming	52.90	11.04	52.58	9.23	0.15	.03	.88
Guilty-dysphoric daydreaming	37.03	7.85	35.76	9.61	0.64	.14	.52
Hostile daydreaming	8.57	2.91	7.01	2.94	2.44	.53	.02
Vivid and colorful night dreams	8.70	1.34	7.51	1.91	3.02	.69	.01
Fantasy scale	22.33	4.74	21.49	4.22	0.89	.19	.38
Composite imagination factor scores	.39	.96	-.16	.87	2.78	.61	.01

Note. d is the effect size for the t test.

TABLE 3
CORRELATIONS AMONG MEASURES OF PERSONALITY

	Shyness scale	Moving toward	Moving away	Moving against	Impostor phenomenon	Hyper-sensitive	Public S-C	Introspective	Internal state
Moving toward others	.31								
Moving away from others	.43	-.01							
Moving against others	-.10	.37	.15						
Impostor phenomenon	.57	.30	.49	.12					
Hypersensitive narcissism	.68	.36	.43	.12	.59				
Public self-consciousness	.37	.42	.07	.05	.37	.42			
Introspectiveness	.29	.27	.07	.02	.28	.37	.56		
Internal state awareness	.14	.28	-.13	.05	.14	.21	.40	.48	
Happy childhood memories	-.11	-.08	-.15	-.30	-.29	-.20	-.05	.05	.14

$N = 102$, for $rs > .20$, $p < .05$

cases. Specifically, participants reporting companions endorsed items related to habitual use of imagery, hostile daydreams, and vivid and colorful night dreams significantly more highly than did participants who did not report pretend friends. The two groups were not significantly different for daydreaming frequency, positive versus guilty daydreams, or on the fantasy scale relating to involvement in fiction. In order to examine all seven measures of imagination in one summary analysis, a weighted and standardized composite measure was created using factor scores from the first unrotated factor of a factor analysis of the imagination measures. The loadings on the first factor ranged from .27 to .81, and the standardized factor scores had a mean of 0 (SD = .93). As may be seen in the last row of Table 2, participants reporting imaginary companions scored significantly higher on this composite measure of imagination than did their counterparts in the other group, suggesting that women who reported imaginary companions made more use of imagery and imagination than women who did not.

Personality

The correlations among the personality scales are presented in Table 3. The measure of the interpersonal style of moving toward others did not correlate significantly with the measures of moving away from or moving against others, but it did have small to moderate positive correlations with most of the other per-

TABLE 4
GROUP DIFFERENCES ON PERSONALITY MEASURES

	Imaginary companion n = 30		No imaginary companion n = 72				
	M	SD	M	SD	t	d	p
Shyness	57.93	15.62	56.53	16.30	.40	.09	.69
Interpersonal styles							
Toward others	76.70	10.67	70.18	11.69	2.63	.57	.01
Away from others	38.93	8.85	39.29	8.82	.19	.04	.85
Against others	31.43	8.88	30.11	7.56	.76	.17	.45
Impostor phenomenon	66.43	13.31	61.54	14.79	1.57	.34	.12
Hypersensitive narcissism	31.33	6.20	29.90	6.10	1.07	.23	.29
Public self-consciousness	27.97	4.12	27.13	4.35	.90	.20	.37
Private self-consciousness							
Introspectiveness	62.90	10.93	62.44	7.82	.24	.05	.81
Internal state awareness	16.50	2.60	14.81	2.48	3.11	.67	.01
Happy childhood memories	114.50	21.42	120.44	17.44	1.46	.32	.15

Note. d is the effect size for the t test.

sonality scales. As may be seen in Table 4, significant group differences emerged for two personality measures: the interpersonal style of moving toward others and the internal state awareness factor of private self-consciousness. In both cases, mean ratings of the imaginary companion group were higher than were those of the group without imaginary companions. No other significant differences emerged on the personality variables, suggesting that women reporting imaginary companions differ from those who do not in their orientation toward pleasing other people and in their monitoring of their own emotional states.

DISCUSSION

The image of the adult woman who recollects an imaginary companion from childhood suggested by the results is of a person with a high imagery use, vivid and colorful night dreams, an orientation toward others, and awareness of her own internal states. This image is largely consistent with the picture painted by the literature on imaginary companions and helps to define those aspects of imagination and personality that distinguish between those who did and did not create imaginary companions.

The differences on measures of imagination suggest that the variation between women with and without childhood pretend friends is partly a function of imaginative skills. In particular, the higher scores of women with imaginary companions on the composite imagination factor demonstrate individual differences in the use of imagination overall. The lack of difference on frequency of daydreaming and involvement in fiction implies that women who recollect imaginary companions do not spend more of their time engaged in fantasy or involved in externally provided fantasy than do their peers. Rather, they differ from their peers in the extent to which they use imagery day-to-day (not necessarily for daydreaming) and in the vividness of their dreams at night. The source of this difference cannot be determined from this study, but in combination with the literature on children who develop imaginary companions, it suggests that these individuals may show a predisposition for using fantasy both early in life and in adulthood.

The content of the day and night dreams of women who do and do not recall imaginary companions also differs. One interpretation for the finding that they engage in hostile daydreams more than do their peers is that they use their imagery in processing their anger. This explanation is also consistent with the findings of Dierker et al. (1995) relating higher scores on measures of normative dissociation to women with vivid memories of imaginary companions over other women. Handling negative emotions using imagery might be a viable coping mechanism for women who dissociate easily as they can do so without being threatened by the content. Conversely, if they are more anxious about social interaction as is suggested by Bonne and colleagues (1999), then playing out hostile interactions in their minds might be their way of reducing anxiety about difficult interactions.

The lack of significant findings on many of the personality variables is consistent with past research on personality differences between those with and without imaginary companions (e.g., Taylor, 1999). In particular, the similarity between the groups on shyness replicates work on children with and without imaginary companions (Partington & Grant, 1984; Singer & Singer, 1990). Notably, the lack of difference in narcissism contradicts Wingfield (1948), who found that women with imaginary companions were less narcissistic than their peers were.

The prediction that personality characteristics related to relationships with others would differ between groups received support from the measure of interactional style referring to moving toward others, but not by either of the other interactional styles or the measures of self-consciousness. Moving toward others included items relating to dependency and pleasing other people, suggesting that women with imaginary companions differ from their counterparts primarily in terms of their willingness to accede to others' wishes for the sake of maintaining the relationship. They appear to place a high value on harmony with others and on sociability, even if it means going against their own desires. Valuing positive interactions in this way seems consistent with Wingfield's (1948) findings regarding the tendencies of women with memories of imaginary companions to dislike solitude, seek advice and encouragement, and be more extroverted than their peers are. Such an interpretation is also consistent with the literature demonstrating a relation between imaginary companion creators and anxiety about social interactions (Bonne et al., 1999). Women who acquiesce to others against their own wishes for the sake of getting along may experience conflict regarding their social interactions.

Taken together with the literature on imaginary companions earlier in life, the findings concerning personality characteristics suggest a developmental pathway that individuals who create imaginary companions may be following. They appear to develop a social orientation early on that includes cooperativeness and easy interactions with adults and little aggression in play (Manosevitz et al., 1973; Singer & Singer, 1990), as well as concerns regarding meeting adults' expectations (Bouldin & Pratt, 2002). Even in preschool these children may be attuned to the needs of others and inclined to be accommodating. In adolescence, this orientation may be translated into a coping style in which the adolescent makes extensive use of her social resources (Seiffge-Krenke, 1997). At both points in time as well as in adulthood, those who create imaginary companions, and women in particular, appear to value harmonious social interaction.

The prediction that women who recollected pretend friends would be more introspective and aware of their own states than were their peers was only partially supported. The pattern of the findings suggest that women recalling imaginary companions do not spend more time analyzing themselves but are more aware than are their peers of their own thoughts and feelings from one moment to the next. Perhaps the attentional focusing characteristic of children with imaginary companions (Singer, 1961) is manifested in adult women partially in terms of a heightened awareness of one's moment-to-moment internal states.

Differences between those with and without imaginary companions, particularly on the personality variables of interaction style and internal state awareness, appear to be stronger than are differences discussed in the childhood literature. No doubt this effect is at least partly a result of the fact that not all adults who had imaginary companions as children remember them. Those who do remember their companions may have had particularly vivid experiences or may be in the extremes of the imaginary companion group in other ways. In addition, we did not ask for descriptions of what participants considered "having an imaginary companion," meaning that the companions reported by the participants could have been different (and perhaps unrelated) types of phenomena. However, some validity for using these memories to define groups is provided by the statistical similarities on childhood memories. The memories of childhood of the two groups did not differ in at least one fundamental way other than by imaginary companion status.

The findings presented here suggest several avenues for future research. First, further examination of women's concepts of relationships and relationship styles might be fruitful for understanding the different orientations of women who did and did not report imaginary companions. Differences in relational orientations due to culture would be of interest given that cultures have distinctly different values concerning individualism and collectivism (Triandis, 1995). Because imaginary companions are part of children's social networks (Gleason, 2002), and social networks function differently according to culture, the need for an imaginary companion could vary across ethnic and cultural lines. The development of social interaction styles would be particularly revealing, as they most likely have their roots in children's earliest interactions. In fact, some evidence suggests that imaginary companions may function similarly to real relationship partners (Gleason, Sebanc, & Hartup, 2000), and these relationships might be informative concerning how children begin to orient themselves with respect to others. Second, the use and type of imagery employed by individuals with and without memories of imaginary companions may be playing a greater role in their daily functioning than has been suspected in the past. Investigation into the precise nature of this imagery and its use might shed light on the different ways that individuals process both social and nonsocial information.

REFERENCES

Acredolo, L., Goodwyn, S., & Fulmer, A. (1995, April). *Why some children create imaginary companions: Clues from infant and toddler play preferences.* Poster session presented at the biennial meetings of the Society for Research in Child Development, Indianapolis, IN.

Ames, L., & Learned, J. (1946). Imaginary companions and related phenomena. *Journal of Genetic Psychology, 69,* 147-167.

Aron, E. N. (1996). *The highly sensitive person: How to thrive when the world overwhelms you.* Secaucus, NJ: Carol Publishing Group.

Aron, E. N., & Aron, A. (1997). Sensory-processing sensitivity and its relation to introversion and emotionality. *Journal of Personality and Social Psychology, 73,* 345-368.

Avila, A. (2002). *The gift of shyness.* NY: Fireside Books.

Bach, L. M., Chang, A. S., & Berk, L. E. (2001, April). *The role of imaginary companions in the development of social skills and play maturity.* Poster presented at the annual meetings of the Eastern Psychological Association, Washington, DC.

Bonne, O., Canetti, L., Bachar, E., DeNour, A.-K., & Shalev, A. (1999). Childhood imaginary companionship and mental health in adolescence. *Child Psychiatry and Human Development, 29,* 277-286.

Bouldin, P., & Pratt, C. (1999). Characteristics of preschool and school-age children with imaginary companions. *Journal of Genetic Psychology, 160,* 397-410.

Bouldin, P., & Pratt, C. (2002). A systematic assessment of the specific fears, anxiety level, and temperament of children with imaginary companions. *Australian Journal of Psychology, 54,* 79-85.

Burnkrant, R. E., & Page, T. J. (1984). A modification of the Fenigstein, Scheier, and Buss Self-Consciousness Scales. *Journal of Personality Assessment, 48,* 629-637.

Buss, A. H., & Perry, M. (1992). The Aggression Questionnaire. *Journal of Personality and Social Psychology, 63,* 452-459.

Caughey, J. (1984). *Imaginary social worlds.* Lincoln, NE: University of Nebraska Press.

Cheek, J. M., & Buss, A. H. (1981). Shyness and sociability. *Journal of Personality and Social Psychology, 41,* 330-339.

Cheek, J. M., & Krasnoperova, E. N. (1999). Varieties of shyness in adolescence and adulthood. In L. A. Schmidt & J. Schulkin (Eds.), *Extreme fear, shyness, and social phobia: Origins, biological mechanisms, and clinical outcomes* (pp. 224-250). New York: Oxford University Press.

Cheek, J. M., & Madaffari, T. M. (1998, August). *Shyness and the many meanings of introversion.* In B.J. Carducci & J.M. Cheek (Co-Chairs), New directions in shyness research and treatment. Symposium conducted at the annual meeting of the American Psychological Association, San Francisco.

Cheek, J. M., & Melchior, L. A. (1985, August). *Measuring the three components of shyness.* In M. H. Davis & S. L. Franzoi (Chairs), Emotion, personality, and personal well-being. Symposium conducted at the meeting of the American Psychological Association, Los Angeles.

Clance, P. (1985). *The impostor phenomenon.* Atlanta: Peachtree Publishers.

Cohen, J. (1988). *Statistical power analysis for the behavioral sciences.* Hillsdale, NJ: Erlbaum.

Davis, M. (1983). Measuring individual differences in empathy: Evidence for a multidimensional approach. *Journal of Personality and Social Psychology, 44,* 113-126.

Dierker, L. C., Davis, K. F., & Sanders, B. (1995). The imaginary companion phenomenon: An analysis of personality correlates and developmental antecedents. *Dissociation: The Official Journal of the International Society for the Study of Multiple Personality and Dissociation, 8,* 220-228.

Fenigstein, A., Scheier, M. F., & Buss, A. H. (1975). Public and private self-consciousness: Assessment and theory. *Journal of Consulting and Clinical Psychology, 43,* 522-527.

Gleason, T. (2002). Social provisions of real and imaginary relationships in early childhood. *Developmental Psychology, 38*, 979-992.

Gleason, T., Sebanc, A., & Hartup, W. (2000). Imaginary companions of preschool children. *Developmental Psychology, 36*, 419-428.

Guilford, J. P. (1959). *Personality.* New York: McGraw-Hill.

Hall, E. (1982). The fearful child's hidden talents [Interview with Jerome Kagan]. *Psychology Today, 16* (July), 50-59.

Hendin, H. M., & Cheek, J. M. (1997). Assessing hypersensitive narcissism: A re-examination of Murray's Narcism Scale. *Journal of Research in Personality, 31*, 588-599.

Hiscock, M. (1978). Imagery assessment through self-report: What do imagery questionnaires measure? *Journal of Consulting and Clinical Psychology, 46*, 223-230.

Horney, K. (1945). *Our inner conflicts.* New York: Norton.

Huba, G. J., Singer, J. L., Aneshensel, C. S., & Antrobus, J. S. (1982). *The Short Imaginal Processes Inventory.* Port Huron, MI: Research Psychologists Press.

Hurlock, E., & Burstein, M. (1932). The imaginary playmate: A questionnaire study. *Journal of Genetic Psychology, 41*, 380-392.

Manosevitz, M., Fling, S., & Prentice, N. (1977). Imaginary companions in young children: Relationships with intelligence, creativity and waiting ability. *Journal of Child Psychology and Psychiatry, 18*, 73-78.

Manosevitz, M., Prentice, N., & Wilson, F. (1973). Individual and family correlates of imaginary companions in preschool children. *Developmental Psychology, 8*, 72-79.

Mauro, J. (1991). The friend that only I can see: A longitudinal investigation of children's imaginary companions (Doctoral dissertation, University of Oregon, Eugene, 1991). *Dissertation Abstracts International, 52*, 4995.

Meyer, J., & Tuber, S. (1989). Intrapsychic and behavioral correlates of the phenomenon of imaginary companions in young children. *Psychoanalytic Psychology, 6*(2), 151-168.

Miller, L. C., Murphy, R., & Buss, A. H. (1981). Consciousness of body: Private and public. *Journal of Personality and Social Psychology, 41*, 397-406.

Nagera, H. (1969). The imaginary companion: Its significance for ego development and conflict solution. *Psychoanalytic Study of the Child, 24*, 165-195.

Paivio, A., & Harshman, R. (1983). Factor analysis of a questionnaire on imagery and verbal habits and skills. *Canadian Journal of Psychology, 37*, 461-483.

Partington, J., & Grant, C. (1984). Imaginary playmates and other useful fantasies. In P. Smith (Ed.), *Play in animals and humans* (pp. 217-240). New York: Basil Blackwell.

Piliavin, J. A., & Charng, H. (1988). What is the factor structure of the private and public self-consciousness scales? *Personality and Social Psychology Bulletin, 14*, 587-595.

Robins, C. J., Ladd, J., Welkowitz, J., Blaney, P. H., Diaz, R., & Kutcher, G. (1994). The Personal Styles Inventory: Preliminary validation of new measures of sociotropy and autonomy. *Journal of Psychopathology and Behavioral Assessment, 16*, 277-300.

Schaefer, C. (1969). Imaginary companions and creative adolescents. *Developmental Psychology, 1*, 747-749.

Scheier, M. F., & Carver, C. S. (1985). The Self-Consciousness Scale: A revised version for use with general populations. *Journal of Applied Social Psychology, 15*, 687-699.

Seiffge-Krenke, I. (1993). Close friendship and imaginary companions in adolescence. *New Directions for Child Development, 60*, 73-87.

Seiffge-Krenke, I. (1997). Imaginary companions in adolescence: sign of a deficient or positive development? *Journal of Adolescence, 20*, 137-154.

Singer, D., & Singer, J. (1990). *The house of make-believe.* Cambridge: Harvard University Press.

Singer, J. (1961). Imagination and waiting ability in young children. *Journal of Personality and Social Psychology, 29,* 396-413.

Singer, J. L., & Antrobus, J. S. (1970). *The Imaginal Processes Inventory.* Princeton, NJ: Educational Testing Service.

Singer, J.L., & Bonanno, G.A. (1990). Personality and private experience: Individual variations in consciousness and in attention to subjective phenomena. In L.A. Pervin (Ed.), *Handbook of personality: Theory and research* (pp. 419-444). New York: Guilford.

Svendsen, M. (1934). Children's imaginary companions. *Archives of Neurology and Psychiatry, 32,* 985-999.

Taylor, M. (1999). *Imaginary companions and the children who create them.* New York: Oxford University Press.

Triandis, H. (1995). *Individualism and collectivism.* Boulder, CO: Westview Press.

Vostrovsky, C. (1895). A study of imaginary companions. *Education, 15,* 393-398.

Waters, P. L., & Cheek, J. M. (in press). Personality development. In V.J. Derlega, B.A. Winstead, & W.H. Jones (Eds.), *Personality: Contemporary theory and research* (3rd ed.). Belmont, CA: Wadsworth.

Watkins, M. (2000). *Invisible guests: The development of imaginal dialogues.* Woodstock, CT: Spring Publications.

Wingfield, R. (1948). Bernreuter personality ratings of college students who recall having had imaginary playmates during childhood. *Journal of Child Psychiatry, 1,* 190-194.

"Influence of seeding depth and seedbed preparation on establishment, growth and yield of fibre flax (Linum usitatissimum L.) in Eastern Canada," Couture, et al, *J. Agronomy & Crop Science*, 190:184–190, 2004.[1]

In the section titled "Introduction" the authors give a good overview of the issues in growing flax. This flax is grown for the fibre (Canadian spelling) used in linen. The article mentions "retted straw." "Retted straw yield" is discussed on p. 185, second column.

1. Tables 1–3 contain similar information for three different sites in 1997. Refer to p. 185 and draw a sketch of the layout for each field. What are the N's for Tables 1–3? That is, in Table 1, the first entry is 476 m^2. How many measurements were averaged to get this value?
2. One of the goals of the study was to find the optimum depth for planting flax. The authors used ANOVA. Why didn't they use linear regression? If they had used regression, how would they determine the optimal depth? How could they determine if the optimum was the same for different rolling conditions?
3. Each of the tables use Tukey's HSD test. "HSD" stands for Honestly Significant Differences. What extra information does this give you that ANOVA does not? You may have covered other methods in your course, like the Bonferroni correction. Why can't we just use a two-sample t test?
4. In the description of Tukey's HSD at the bottom of the tables, there is the notation "$p \geq 0.05$." We don't usually see $\geq$ in tables of statistics. Why does it appear here?
5. For all such tests, we need an estimate of the standard deviation of the data. Where would we get this estimate from the design used here?
6. Tukey's HSD is displayed using letters to indicate values that are not different from each other. In Table 1, only the column for retted straw has these letters. Why is that?
7. In Table 1, under Not rolled, two of the values are marked with "c." Why aren't these two values adjacent to each other? The first value (2.2) is the only one marked with "d." What does this indicate?
8. At the bottom of Table 1 is a row labeled "m.s.d." What does this signify? Why is this value present only for retted straw? For which of the "rolled" levels does this m.s.d apply?
9. Tables 2 and 3 use Tukey's HSD for columns other than retted straw. Why do they do this when they didn't do it for Table 1?
10. In Table 3, under Branching ratio, all of the letters for Tukey's HSD test are "b" except for a single "a." What does this mean? Does this make sense? If you were using Table 3 as a guide for planting flax, what would you think of the data on Branching ratio?
11. On p. 187, the authors say that there is a significant interaction between Treatment and Depth for Height in Table 3. How does what they say in the first paragraph on p. 187 support their claim of interaction? What statistic did they use to back up this claim? Where would we find this statistic?

---

[1] © 2004, Blackwell Publishing. Used by permission.

12. The authors note that the tallest plants in Table 3 for Not rolled were those planted at 4 cm. How does this compare to the results from Tukey's HSD for Height? They also note that the shortest plants were under Rolled before and 6 cm. What about Tukey's HSD for this group?
13. Table 4 has some columns that do not appear in Tables 1 - 3. Why do you think this is? What is the other difference between Tables 1 - 3 and Table 4?
14. All the depths in Table 4 have an additional annotation of NR, RA, or RB. What additional information is conveyed by these annotations?

Department of Crop and Soil Sciences, Cornell University, Ithaca, NY, USA

Influence of Seeding Depth and Seedbed Preparation on Establishment, Growth and Yield of Fibre Flax (*Linum usitatissimum* L.) in Eastern Canada

S. J. Couture, A. DiTommaso, W. L. Asbil, and A. K. Watson

Authors' addresses: Mr S. J. Couture (present address: Herbicides and PGRs Section, Pest Management Regulatory Agency, Health Canada, 2720 Riverside Drive, Ottawa, Ontario, Canada K1V 7V1), Dr A. DiTommaso (corresponding author; e-mail: ad97@cornell.edu; present address: Department of Crop and Soil Sciences, Cornell University, Ithaca, NY 14853, USA) and Dr A. K. Watson, Department of Plant Science, Macdonald Campus, McGill University, 21,111 Lakeshore Road, Ste. Anne-de-Bellevue, Québec, Canada H9X 3V9; Ms W. L. Asbil, Kemptville College, University of Guelph, Kemptville, Ontario, Canada K0G 1J0

With 4 tables

Received September 10, 2003; accepted November 18, 2003

Abstract

Research was conducted at the Macdonald Campus of McGill University (Québec, Canada) at three sites in 1997 and one site in 1998 to determine the effects and interactions of seeding depth (0, 1, 2, 4 or 6 cm) and seedbed preparation (i.e. soil rolling): none, rolling before or rolling after seeding on fibre flax (cv. Ariane) establishment, growth and yield. Seedbed preparation had little impact on the parameters measured while seeding depth had a variable effect on plant density, plant height, stem diameter and retted straw yield. Seeding depths of 1–4 cm provided consistently good establishment, growth and yield results. In 1997, there was an interaction between seeding depth and seedbed preparation on plant height, branching ratio and retted straw yield, although results were generally variable and tended to be site-specific. In 1998, there was an interaction between seeding depth and seedbed preparation on plant height and stem diameter prior to harvest, with the results varying for all seeding depth-seedbed preparation treatment combinations except for the 2-cm depth treatment. Rolling of the seedbed before seeding on lighter soils and at a depth of 2 cm on most soils can improve establishment, growth and yields of fibre flax under eastern Canadian growing conditions.

Key words: Canada—fibre flax—*Linum usitatissimum*—seedbed preparation—seeding depth—soil rolling

Introduction

Early and uniform establishment is paramount to the success of fibre flax crops for a number of reasons. Fibre flax is seeded at very high densities to attain optimal populations of 2000 plants m^{-2} (Sultana 1983, Stephens 1997), so even a partial delay in emergence can result in highly non-uniform stands as plants emerging late are shorter and at a competitive disadvantage (Fowler 1984). Also, uniformity of plant height is a desirable characteristic in fibre flax destined for linen production because the longer and more uniform the fibres; the more valuable is the crop (Ulrich and Laugier 1995).

Stand uniformity is influenced by seed placement and seedbed preparation (Lafond et al. 1996). Seed placement plays a major role in the time to emergence of flax seedlings and may also impact seedling vigour. Sultana (1983) reported an optimal depth of 2 cm for the sowing of fibre flax. Work by O'Connor and Gusta (1994) on oilseed flax showed that flax sown at a depth of 4 cm required 33 % longer to emerge than flax sown at a depth of 2 cm, and that overall emergence was lower from a depth of 4 cm. Wall (1994) reported reductions of up to 59 % in oilseed flax populations when seeded at depths of 6 cm vs. 3 cm.

One practice commonly used in forage production to improve seedling emergence is the compacting or rolling of the soil before or after seeding as a way to improve seed-to-soil contact and seedbed firmness. This practice is also likely to impact the emergence and early growth of other crop species (Lafond et al. 1996). Results of a study in Québec (Canada) by Robert (1998) suggest that seeding depth and seedbed preparation may be two important factors influencing fibre flax production. For instance, seeds placed too deep in the soil may emerge in a discontinuous manner leading to uneven crop development and a high proportion of immature plants at harvest. Similarly, a seedbed that is not firm, especially on lighter soils, may also lead to uneven crop emergence and stand development because of poor seed-to-soil contact (Robert 1998).

Hence, the objective of this study was to examine the effects and possible interactions of seeding depth and soil seedbed preparation (rolling) before seeding, after seeding, or not at all, on fibre flax (cv. Ariane) establishment, growth, and yield in a conventional tillage system in Québec (Canada). Although this type of research has been well documented in other parts of the world, this study is part of a larger project examining the feasibility of re-introducing fibre flax production in Canada.

Materials and Methods

The research was conducted in 1997 and 1998 at the Emile A. Lods Agronomy Research Centre of Macdonald Campus (McGill University, Ste-Anne-de-Bellevue, Québec, Canada) (43°25′N, 73°56′W). There were three experimental sites in 1997 and one site in 1998. The first site was located on a Bearbrook clay soil (poorly drained, Dark Gray Gleysolic; pH 5.6–6.4) (Lajoie 1960), the second site was located on a St Amable loamy sand soil (well drained, Gleyified Humo Ferric Podzol; pH 5.0–5.7) (Lajoie 1960), and the third site was located on Macdonald clay loam soil (poorly drained, Dark Gray Gleysolic; pH 5.8–6.2) (Lajoie 1960). Previous crops in each location were wheat in 1996 and maize in 1995 on the Bearbrook clay site, and the Macdonald clay loam site. At the St Amable site, soybean was sown in 1996 and was in fallow in 1995. The single site in 1998 was located on a St Bernard loam soil (well drained, Melanic Brunisol, pH 6.0–7.0) (Lajoie 1960), and previous crops on this site were barley in 1997, and red clover in 1995 and 1996.

Land preparation in both years consisted of fall mouldboard plowing followed by spring disking and harrowing prior to seeding. In both years, 200 kg ha^{-1} of balanced inorganic fertilizer (20–10–10) (N–P$_2$O$_5$–K$_2$O) was broadcast and incorporated prior to seeding.

The fibre flax cultivar 'Ariane' was seeded at rates of 100 (1997) and 125 (1998) kg ha^{-1} using a five row Bolens drill plot seeder having a row spacing of 20 cm, at manually controlled depths of 0, 1, 2, 4 and 6 cm, as measured in the field. Seeding occurred on 25 May in 1997 and 30 April in 1998 into 2 m × 5.5 m plots. The experimental design was a strip-plot in which the strips (vertical factor) comprised the seeding depth treatment, and the blocks (horizontal factor) comprised one of three seedbed preparation treatments: (i) rolling prior to seeding (RB), (ii) rolling after seeding (RA) or (iii) no rolling (NR). There were three blocks, with one replicate/treatment/block. The field layout in all sites in both years was identical.

High weed pressure necessitated herbicide application in both years. On 25 June 1997, the Bearbrook and St Amable clay sites were sprayed with bentazon (BASF Canada Inc., London, Ont., Canada) at a rate of 0.96 kg a.i. ha^{-1} for the control of broadleaf weeds and yellow nutsedge (*Cyperus esculentus* L.). The Macdonald clay loam site was sprayed with sethoxydim (BASF Canada Inc.) + mineral oil surfactant (0.276 kg ai ha^{-1} + 2 l ha^{-1}) for control of annual grasses. The 1998 site was sprayed on 29 May with fluazifop-*p*-butyl (Syngenta Crop Protection Inc., Guelph, Ont., Canada) + mineral oil surfactant (0.7 kg ai ha^{-1} + 0.5 % v/v) for annual grass and quackgrass [*Elytrigia repens* (L.) Nevski] control. Bentazon was applied on 5 June (1.08 kg ai ha^{-1}) on this site to control emerged annual broadleaf weeds. All herbicide treatments followed recommendations for oilseed flax in Ontario (Anonymous 1997).

Data collection in both years included mean plant height per plot and was recorded approximately 9 weeks post-emergence. Height was determined by measuring plants from soil level to the uppermost growing point. The mean height per plot was the average of two observations in 1997 and six observations in 1998. At each of these sampling times, 8–12 plants were held together against a meter-stick, and their mean height estimated.

Plant densities in each plot were assessed 42 days after seeding (DAS) within four randomly placed 0.25-m^2 quadrats. The number of plants with branches on the lower 50 cm of stem was also recorded. A 'branching ratio' (Tratio) value was determined by dividing the number of plants with branches on the lower 50 cm of stem by the number of plants in each 0.25-m^2 quadrat. Stem diameters were measured in 1998 prior to harvest on 25 randomly selected plants per plot using a digital calliper (Marathon Electronics, Belleville, Ont., Canada) and plot mean values were used for data analysis.

'Retted' straw yield was also assessed in each plot. Plants within a 1-m^2 sub-plot were uprooted by hand when at least two thirds of leaves had senesced and capsules were turning brown (Stephens 1997). The soil was shaken from roots, plants were laid on the ground and left in the field for about 2 weeks, after which, they were turned over by hand, and left in the field for 10 additional days to complete the retting process. Retted plants in each plot were then collected and weighed. Due in part to the subjectivity involved in assessing the degree of retting for each treatment independently, fresh and dry weights rather were obtained in 1998 than retted weight. Dry weights were obtained by placing samples in a forced air electric drying unit at 65 °C for 48 h.

All data were subjected to analysis of variance (ANOVA) using the GLM procedure in SAS (SAS Institute Inc. 1985) to identify effects and interactions of seeding depth and seedbed preparation treatments. Treatment mean values were separated using the Tukey's HSD mean values comparison procedure at the P < 0.05 level of significance (Motulsky 1995). Branching ratio and plant density data were subjected to a square root + 1-transformation before analysis to improve the normality requirement of ANOVA (Gomez and Gomez 1984).

Results

1997 Field trials

The growing season in 1997 was much wetter than usual, with only June receiving normal precipitation. Temperatures were below average in April and May, and average for the rest of the season. The season-long number of growing degree-days (GDD) (base 5 °C) was 1828, 143 GDD lower than the 20-year average of 1971.

The effect of seedbed preparation (rolling) was not significant ($P \geq 0.05$) at the Bearbrook site. There was however, a significant ($P < 0.05$) effect of seeding depth on plant density. The 6-cm seeding depth treatment resulted in the highest plant densities (1348 plants m^{-2}), whereas the 0-cm seeding depth treatment resulted in the lowest plant densities (752 plants m^{-2}). There was a highly significant interaction ($P < 0.01$) between seeding depth and seedbed preparation on retted straw yield. The highest yielding treatment combination (6.1 t ha^{-1}) was sowing at a depth of 1 cm with no rolling, while sowing at a 0-cm depth and rolling prior to seeding resulted in no measurable yields (Table 1). The seeding depth and seedbed preparation treatments had no effect on the branching ratio and height of flax plants.

Results from the St Amable site were similar to those obtained at the Bearbrook clay site in that there was no main effect of the seedbed preparation treatment on any of the parameters measured. However, seeding depth had a significant effect on plant density. The 1-cm sowing depth treatment resulted in the highest plant density (1089 plants m^{-2}) while sowing at a depth of 6 cm resulted in the lowest plant density (516 plants m^{-2}). There was a significant interaction ($P < 0.05$) between the seeding depth and seedbed preparation treatments on branching ratio and retted straw yield. The combination of the 0-cm seeding depth treatment and rolling prior to seeding produced the highest proportion (0.46) of branched plants (Table 2). The combination of the 1-cm seeding depth and rolling prior to seeding treatments yielded the most retted straw at 8.3 t ha^{-1} (Table 2). In contrast, the 4- and 6-cm sowing depth treatments that were rolled after seeding yielded only 3.0 and 2.7 t ha^{-1} of retted straw, respectively (Table 2).

Table 1: Comparison of fibre flax parameter mean values for all seeding depth-seedbed preparation treatment combinations at the Bearbrook clay site in 1997

Treatment	Plant density (m^2)	Branching ratio	Retted straw yield (t ha^{-1})	Height (cm)
Not rolled				
0 cm	476	0.28	2.2 cd	73.3
1 cm	1564	0.12	6.1 a	78.7
2 cm	1052	0.19	5.7 ab	79.7
4 cm	972	0.09	4.0 abc	71.7
6 cm	1636	0.10	5.3 ab	77.0
Rolled after				
0 cm	912	0.19	2.8 bcd	77.7
1 cm	1208	0.16	4.6 abc	77.0
2 cm	1124	0.10	5.6 ab	83.7
4 cm	1000	0.11	4.0 abc	78.7
6 cm	1104	0.12	4.6 abc	73.0
Rolled before				
0 cm	868	0.19	0 d	70.3
1 cm	1060	0.12	4.8 abc	74.3
2 cm	876	0.36	5.2 abc	74.7
4 cm	844	0.18	3.9 abc	73.3
6 cm	1304	0.12	5.3 ab	76.0
m.s.d.	–	–	3.1	–

Values within a column followed by the same letters do not differ, according to Tukey's HSD test ($P \geq 0.05$).
NR, no rolling; RA, rolled after seeding; RB, rolled before seeding.
m.s.d., minimum significant difference.

Table 2: Comparison of fibre flax parameter mean values for all seeding depth-seedbed preparation treatment combinations for the St Amable sandy loam site in 1997

Treatment	Plant density (m^2)	Branching ratio	Retted straw yield (t ha^{-1})	Height (cm)
Not rolled				
0 cm	808	0.17 ab	5.7 ab	88.0
1 cm	852	0.15 ab	6.7 ab	90.7
2 cm	784	0.15 ab	5.8 ab	92.0
4 cm	656	0.03 b	5.8 ab	87.0
6 cm	712	0.13 ab	3.8 ab	89.7
Rolled after				
0 cm	1120	0.09 b	5.8 ab	87.7
1 cm	1008	0.21 ab	6.5 ab	92.0
2 cm	880	0.07 b	5.4 ab	92.0
4 cm	756	0.14 ab	3.0 b	88.7
6 cm	296	0.12 ab	2.7 b	86.7
Rolled before				
0 cm	880	0.46 a	4.2 ab	88.3
1 cm	1408	0.05 b	8.3 a	83.3
2 cm	1000	0.16 ab	7.5 ab	88.7
4 cm	852	0.09 b	6.1 ab	90.0
6 cm	540	0.25 ab	5.5 ab	87.0
m.s.d.	–	0.36	5.1	–

Values within a column followed by the same letters do not differ, according to Tukey's HSD test ($P \geq 0.05$).
NR, no rolling; RA, rolled after seeding; RB, rolled before seeding.
m.s.d., minimum significant difference.

At the Macdonald clay loam site, there were significant ($P < 0.01$) interactions between seeding depth and seedbed preparation on plant height, branching ratio and retted straw yield. The tallest plants (81.0 cm) were found in the 4-cm sowing depth–non-rolled treatment combination plots, while the shortest plants (63.7 cm) were found in the 6-cm sowing depth plots rolled prior to seeding (Table 3). Plants with the highest branching ratio (0.62) were found in plots sown to a depth of 0 cm and rolled prior to seeding (Table 3). Plants did not differ significantly in branching ratio for any of the other treatment combinations. Sowing at a depth of 4 cm and rolling the soil prior to seeding resulted in the highest yield of retted straw (6.3 t ha^{-1}) while sowing at a depth of 0 cm and rolling the soil prior to seeding resulted in the lowest yield of retted straw (1.5 t ha^{-1}) (Table 3). Plant densities of flax were not affected by either the seeding depth or seedbed preparation treatments.

1998 Field trials

April 1998 was considerably warmer and drier than the 20-year average for this month. The mean temperature was 2 °C above normal, with only a quarter of the normal rainfall, thus allowing seeding to take place relatively early. Temperatures were above normal from April to September, except for July, which was 1 °C below normal. The same pattern was observed for rainfall, with every month having below normal precipitation, except for June, which had 1.5 times the normal rainfall. The number of GDD (base 5 °C) was 2335, 364 GDD greater than the 20-year average of 1971.

There was a highly significant ($P < 0.01$) interaction between seeding depth and seedbed preparation on stem diameter and mean plant height 9 weeks following emergence. Stem diameters were greatest for plants in plots seeded at a depth of 6 cm and rolled after seeding and for plants in the 0-cm depth treatment rolled prior to seeding (Table 4). Stem diameters were lowest for plants in the 1- and 2-cm sowing depth plots rolled after seeding, and for the 2-cm sowing depth plots either rolled prior to seeding or not rolled at all (Table 4). Overall, plants within the 0-cm depth plots with no rolling were tallest (92.4 cm) while plants within the 2-cm depth treatment plots rolled prior to seeding were shortest (84.4 cm) (Table 4). Seeding depth and seedbed preparation treatments had no effect

Table 3: Comparison of fibre flax parameter mean values for all seeding depth-seedbed preparation treatment combinations for the Macdonald clay loam site in 1997

Treatment	Plant density (m^2)	Branching ratio	Retted straw yield (t ha^{-1})	Height (cm)
Not rolled				
0 cm	1348	0.14 b	5.7 a	75.0 abc
1 cm	888	0.25 b	5.7 a	75.0 abc
2 cm	1356	0.12 b	3.3 ab	78.3 ab
4 cm	1400	0.12 b	5.3 a	81.0 a
6 cm	1156	0.13 b	5.9 a	79.7 ab
Rolled after				
0 cm	1224	0.18 b	5.8 a	73.3 abcd
1 cm	1888	0.11 b	5.1 ab	67.3 cd
2 cm	1644	0.09 b	4.9 ab	72.7 abcd
4 cm	1556	0.23 b	4.6 ab	67.0 cd
6 cm	1828	0.19 b	4.6 ab	76.3 abc
Rolled before				
0 cm	920	0.62 a	1.5 b	72.0 abcd
1 cm	1624	0.17 b	5.4 ab	73.3 abcd
2 cm	2096	0.10 b	5.5 a	75.0 abc
4 cm	1556	0.15 b	6.3 a	70.3 bcd
6 cm	1468	0.07 b	6.2 a	63.7 d
m.s.d.	–	0.34	3.7	10.2

Values within a column followed by the same letters do not differ, according to Tukey's HSD test ($P \geq 0.05$).
NR, no rolling; RA, rolled after seeding; RB, rolled before seeding.
m.s.d., minimum significant difference.

Table 4: Comparison of fibre flax parameter mean values for all sowing depth-seedbed preparation treatment combinations for the St Bernard loam soil site in 1998

Treatment	Stem diameter (mm)	Plant density (m^2)	Branching ratio	Height (cm)	Fresh yield (t ha^{-1})	Dry matter content (%)
Not rolled						
0 cm, NR	2.77 ab	1299	0.078	92.4 a	28.5	34.2
1 cm, NR	2.70 ab	1431	0.016	86.3 abc	30.8	33.3
2 cm, NR	2.42 b	1392	0.021	88.2 abc	30.3	34.5
4 cm, NR	2.74 ab	1478	0.026	85.8 bc	27.1	38.3
6 cm, NR	2.62 ab	1229	0.029	85.7 abc	24.0	38.7
Rolled after						
0 cm, RA	2.75 ab	1408	0.094	88.2 abc	32.8	34.8
1 cm, RA	2.44 b	1634	0.018	86.7 abc	28.8	37.3
2 cm, RA	2.42 b	1424	0.031	87.1 abc	24.2	35.3
4 cm, RA	2.74 ab	1525	0.049	86.4 abc	28.9	34.7
6 cm, RA	3.12 a	1516	0.067	88.3 abc	25.0	38.0
Rolled before						
0 cm, RB	3.10 a	1579	0.071	90.7 ab	26.5	35.2
1 cm, RB	2.51 ab	1540	0.078	86.6 abc	25.8	37.3
2 cm, RB	2.42 b	1376	0.080	84.4 c	26.1	37.2
4 cm, RB	2.59 ab	1525	0.072	88.9 abc	29.4	37.7
6 cm, RB	2.59 ab	1275	0.106	86.8 abc	29.4	37.0
m.s.d.	0.65	–	–	6.1	–	–

Values within a column followed by the same letters do not differ, according to Tukey's HSD test ($P \geq 0.05$).
NR, no rolling; RA, rolled after seeding; RB, rolled before seeding.
m.s.d., minimum significant difference.

on flax planting density, branching ratio, and fresh and dry yield at this site.

Discussion

Seeding depth is considered to be a major determinant of crop establishment (O'Connor and Gusta 1994), and our results using fibre flax are consistent with this view. A seeding depth of 2 cm was suggested by Sultana (1983) to be optimal for fibre flax, while seeding depths > 4 cm can lead to substantially lower plant densities (Robert 1998). In our trials, a seeding depth of 2 cm generally resulted in average-to-above average fibre flax plant densities, mean heights and yields, and a lower proportion of branched plants. These plant features are considered important for fibre flax production where the fibres are destined for linen production (Hocking et al. 1987). Seeding depths of 1- and 4-cm also provided acceptable levels of establishment, growth and yield.

Seed-to-soil contact is essential in agricultural production systems where typically high levels of germination and emergence of crops is desirable (Lafond et al. 1996). However, findings in this study revealed no effect of seedbed preparation (i.e. rolling) on fibre flax growth and yield for any of the sites in either year, with the exception of plant height at the Macdonald clay loam site in 1997. In some instances, the seedbed preparation treatment did interact with seeding depth, but no clear trends emerged. Despite a lack of significant effects, rolling the plots after seeding generally resulted in greater fibre flax densities in lighter soils (e.g. St Amable site) compared with heavier soils (e.g. Bearbrook site). Unfortunately, because of significant within-site variability as determined by Bartlett's test, we were unable to verify this effect statistically (Gomez and Gomez 1984). In a Québec study, Robert (1998) observed beneficial effects of rolling field plots prior to seeding on fibre flax stand density and retted yields. Unfortunately, the Robert (1998) study did not include a rolling after seeding treatment with which to compare our results. Nonetheless, Robert (1998) did note that when the seedbed was too loose, actual seeding depths were greater than desired and resulted in lower seedling emergence levels than expected. Clearly, under these soil conditions rolling prior to seeding will be most beneficial.

In the 2 years of this study, stand densities never attained optimal levels of 1800–2000 plants m^{-2}. Increasing the seeding rate in 1998 by 25 kg ha^{-1} was beneficial, but flax populations remained at densities nearly 25–30 % below optimal levels. Similarly, Robert (1998) and Couture et al. (2002) also experienced difficulties in achieving optimal population densities of fibre flax under Québec field conditions. In fact, Robert (1997) reported an average density of 900 plants m^{-2} at 35 farms growing fibre flax in Québec. Nonetheless, our flax stand densities appear to be well in line with those reported recently by Rossini and Casa (2003) in field trials in Italy. Although these workers reported that their target density was 2200 plants m^{-2}, they were only able to achieve stand densities at harvest of 1250–1850 plants m^{-2} in 2 of 3 years, and considerably lower densities in last year of the study. Thus, Rossini and Casa (2003) achieved harvest densities of approximately 70 % of the target seeding density, which compares favourably with the 71 % target seeding density achieved in our study. Similarly, Sankari (2000) in Finland reported flax densities of 536–699 plants m^{-2} based on a seeding rate of 800 seeds m^{-2}, for a final stand density of 74 % of the target density.

Although retted straw yields in 1997 appear excessively high relative to the plant densities achieved, exceptionally high rates of branching and increased plant stem diameters might account for this finding. Branching ratios were much reduced in 1998, likely because of the higher seeding rate and increased intra-specific competition amongst plants (Hocking et al. 1987). Contrary to findings by Rowland (1980), plant height in 1998 was not adversely affected by the generally greater plant densities obtained following increases in the seeding rate. In fact, mean plant height actually increased in 1998 compared with 1997.

In conclusion, the impact of seeding depth and seedbed preparation on fibre flax establishment, growth and yield may largely depend on soil type and the specific type of production system used. Our results demonstrate that the rolling of soil prior to seeding on lighter soils and at a seeding depth of 2 cm on most soils can be beneficial for fibre flax production under Québec growing conditions.

Acknowledgements

This work was supported by a research grant from the Conseil des Recherches en Pêche et en Agroalimentaire du Québec (CORPAQ) dans le cadre du Programme de Recherche de L'Entente auxiliare Canada-Québec sur le développement agroalimentaire. The authors wish to thank the Coopérative Linière Fontaine-Cany for supplying much of the seed used in these trials.

References

Anonymous, 1997: Ontario Guide to Weed Control 1998. Publication 75, Ontario Ministry of Agriculture, Food, and Rural Affairs. Queen's Printer (Ontario), Toronto.

Couture, S. J., W. L. Asbil, A. DiTommaso, and A. K. Watson, 2002: Comparison of European fibre flax (*Linum usitatissimum*) cultivars under Eastern Canadian growing conditions. J. Agron. Crop Sci. **188**, 350—356.

Fowler, N. L., 1984: The role of germination date, spatial arrangement, and neighbourhood effects in competitive interactions in *Linum*. J. Ecol. **72**, 307—318.

Gomez, K. A., and A. A. Gomez, 1984: Statistical Procedures for Agricultural Research, 2nd edn. John Wiley & Sons Inc, New York.

Hocking, P. J., P. J. Randall, and A. Pinkerton, 1987: Mineral nutrition of linseed and fiber flax. Adv. Agron. **41**, 221—290.

Lafond, G. P., S. M. Boyetchko, S. A. Brandt, G. W. Clayton, and M. H. Entz, 1996: Influences of changing tillage practices on crop production. Can. J. Plant Sci. **76**, 641—649.

Lajoie, P. G., 1960: Soil Survey of Argenteuil, Two Mountains and Terrebone Counties, Québec. Research Branch, Canada Department of Agriculture in cooperation with Québec Department of Agriculture and Macdonald College, McGill University, Montréal.

Motulsky, H., 1995: Intuitive Biostatistics. Oxford University Press, New York.

O'Connor, B. J., and L. V. Gusta, 1994: Effect of low temperature and seeding depth on the germination and emergence of seven flax (*Linum usitatissimum* L.) cultivars. Can. J. Plant Sci. **74**, 247—253.

Robert, L., 1997: Rapport Annuel 1996 sur la Production de Lin Textile. Ministère de l'Agriculture, des Pêcheries et de l'Alimentation du Québec. Sainte-Martine.

Robert, L., 1998: Rapport Final sur la Production de Lin Textile 1995–97. Ministère de l'Agriculture, des Pêcheries et de l'Alimentation du Québec. Sainte-Martine.

Rossini, F., and R. Casa, 2003: Influence of sowing and harvest time on fibre flax (*Linum usitatissimum* L.) in the Mediterranean environment. J. Agron. Crop Sci. **189**, 191—196.

Rowland, G. G., 1980: An agronomic evaluation of fibre flax in Saskatchewan. Can. J. Plant Sci. **60**, 55—59.

Sankari, H. S., 2000: Linseed (*Linum usitatissimum* L.) cultivars and breeding lines as stem biomass producers. J. Agron. Crop Sci. **184**, 225—231.

SAS Institute Inc. 1985: SAS User's Guide: Statistics, 5th edn. SAS Institute Inc., Cary.

Stephens, G. R., 1997: A Manual for Fiber Flax Production. The Connecticut Agricultural Experiment Station, New Haven, CT.

Sultana, C., 1983: The cultivation of fibre flax. Outlook Agric. **12**, 104—110.

Ulrich, A., and E. Laugier, 1995: Cottonized Flax Fiber: Preliminary Marketing and Irrigated Production Study. Agricultural Development Fund Project no. 94000051, Saskatchewan.

Wall, D. A., 1994: Response of flax and lentil to seeding rates, depths, and spring application of dinitroaniline herbicides. Can. J. Plant Sci. **74**, 875—882.

"Shape of glass and amount of alcohol poured: comparative study of practice and concentration," Brian Wansink and Koert van Ittersum, *British Medical Journal*, 331:1512–1514, 2005.[1]

Note that, although both authors are American, the journal is British. This explains some of the spelling differences and also why, on p. 1513, the amount of money paid to the bartenders is given in dollar, pounds, and euros.

1. Table 1 summarizes the results for the college students. What do the labels "1 trial" and "10 trials" mean in the context of this study? What is the important comparison between these two?
2. The right hand part of Table 1 is labeled "Significance." It has three pairs of columns. What does the pair labeled "Glass shape" signify? What about the pair of columns labeled "Experience"?
3. You may not have covered the notion of interaction. This is what the last pair of columns in Table 1 refers to. In prescription drugs, there is an interaction if taking one drug alters (often enhances) the effect of another drug. How would the notion of interaction apply to "Glass shape" and "Experience"? What about the entries in the last pair of columns? How did the authors make use of this in their analysis?
4. How many students are represented in the first column of Table 1, i.e., "1 trial" and "Tall slender glass"?
5. One thing the authors do not do is to compare the amount poured to the target amount (44.3 ml). How sure are we that the students in "1 trial" and "Tall slender glass" poured more than the target amount? Should we do a one-sided or two-sided test? We can be 95% sure that the mean excess amount is at least how much? (You may not have covered "one-sided confidence intervals. If so, use a "two-sided" interval.) About what fraction of students will "over pour" (by any amount)?
6. The second part of this student involved bar tenders. Under "Attention to pouring alcohol," they report that 86 of 95 bartenders agreed to participate in the study. They note that 62% of those asked were men and 62% of those who agreed to participate were men. What is the importance of this?
7. In the last paragraph under "Attention to pouring alcohol," the authors say that they used "repeated measures analysis of variance." This is different from the usual ANOVA. Find out how it is different and why they needed to use this analysis for their data.
8. Table 2 is somewhat similar to Table 1. What are the main differences in the layouts of the two tables? What is the reason for this difference? What do the authors make of the different rows in Table 2?
9. One important difference between Table 1 and Table 2 is in the last pair of columns. What is the implication of this for the bartender part of the study?
10. On p. 1513 under Results, they compare the amounts poured under different conditions. Do they use confidence intervals or significance testing to do this comparison? What would be the advantage of using the other method?

---

[1] © 2005, BMJ Publishing Group. Used by permission.

11. Under "Attention to pouring alcohol," the time to pour was compared between the high attention group and the low attention group. Where is this data presented? What additional information would you need to confirm their results?
12. In the same section, the paper says "and they agreed more strongly with the statement that they had paid close attention." Agreed "more strongly" than whom?
13. On p. 1514, the paper says that bartenders pour more alcohol into short glasses than into tall ones. What test was used to determine this? Can we determine how much more alcohol?

Shape of glass and amount of alcohol poured: comparative study of effect of practice and concentration

Brian Wansink, Koert van Ittersum

Abstract

Objective To determine whether people pour different amounts into short, wide glasses than into tall, slender ones.

Design College students practised pouring alcohol into a standard glass before pouring into larger glasses; bartenders poured alcohol for four mixed drinks either with no instructions or after being told to take their time.

Setting University town and large city, United States.

Participants 198 college students and 86 bartenders.

Main outcome measures Volume of alcohol poured into short, wide and tall, slender glasses.

Results Aiming to pour a "shot" of alcohol (1.5 ounces, 44.3 ml), both students and bartenders poured more into short, wide glasses than into tall slender glasses (46.1 ml v 44.7 ml and 54.6 ml v 46.4 ml, respectively). Practice reduced the tendency to overpour, but not for short, wide glasses. Despite an average of six years of experience, bartenders poured 20.5% more into short, wide glasses than tall, slender ones; paying careful attention reduced but did not eliminate the effect.

Conclusions To avoid overpouring, use tall, narrow glasses or ones on which the alcohol level is premarked. To avoid underestimating the amount of alcohol consumed, studies using self reports of standard drinks should ask about the shape of the glass.

Introduction

Variations in pouring and drinking behaviour mean that the amount of alcohol consumed from a mixed drink can vary widely.[1-4] Although correction efforts have been suggested,[5 6] an important unaccounted source of bias in self reported consumption of spirits may have to do with the shape of the glass into which a drink is poured.

Two of the most common shapes of glasses for spirits are elongated "highball" glasses and short, wide "tumblers." In one study, adults poured 28% more breakfast juice into short, wide glasses than into slender ones holding the same volume.[7] This is the result of two perceptual biases: people generally estimate that tall glasses hold more liquid than wide ones of the same volume,[8 9] and they focus their pour-

ing attention on the height the liquid reaches and insufficiently compensate for the width of the glass.[7]

Suppose a person wanted to pour a target volume of alcohol, such as a 44.3 ml (1.5 ounce) "shot." The perceptual bias caused by this interaction of vertical and horizontal dimensions could lead to unknowingly pouring more alcohol into a short, wide glass than into a tall, slender glass.

Because people generally consume most (about 92%) of what they have served themselves,[10] this issue of pouring accuracy is relevant to policy makers, health professionals, responsible consumers, law enforcement, and those interested in alcohol addiction and misuse. We examined whether practice in pouring or whether increased concentration can help reduce this potential bias.

Methods

Practice in pouring alcohol

We recruited 198 students of legal drinking age from the University of Illinois at Urbana-Champaign (57% men) through courses in various faculties. They were given partial course credit for their involvement in the study, which had been approved by the university.

A 2×2 between subjects design manipulating shape of glass (short and wide v tall and slender) and pouring education and practice (low v high) was examined across four different drink replications. As participants arrived at the study, they were alternately assigned to

Table 1 Shape of glass and amount of alcohol poured by college students after one or 10 trial pours

| | Mean (SD) amount (ml) | | | | | | Significance | | | | | |
| | Tall, slender glass | | | Short, wide glass | | | Glass shape | | Experience | | Glass shape × experience | |
Variable	1 trial	10 trials	Average	1 trial	10 trials	Average	F value	P value	F value	P value	F value	P value
Perceived capacity of glass	356.5 (221.1)	336.4 (145.1)	346.7	333.7 (137.1)	325.8 (138.5)	329.9	5.46	<0.01	0.64	0.59	0.51	0.68
Volume poured:												
Actual	48.9 (16.2)	42.2 (13.3)	45.5	60.9 (17.9)	57.3 (18.0)	59.6	31.89	<0.01	4.08	<0.01	0.36	0.78
Perceived	46.3 (3.8)	45.9 (2.9)	46.1	44.7 (4.2)	44.6 (4.2)	44.6	7.03	<0.01	0.38	0.77	0.25	0.88

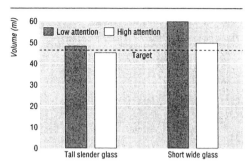

Pouring by bartenders into glasses of different shapes

one of the practice conditions. In the low pouring education and practice condition, participants conducted one practice pour into a 1.5 ounce shot glass, after which the pouring for the experiment began. Participants in the high education condition were asked to conduct 10 practice pours before beginning the pouring for the experiment.

Participants were supplied with full 1500 ml rum and whiskey bottles that had been refilled with brown tea and with 1500 ml gin and vodka bottles that had been refilled with water. Half of the participants were given tall, slender 355 ml glasses and half were given short, wide 355 ml glasses. Participants were asked to pour the amount of liquor that would go into four mixed drinks that were popular at the college—vodka tonic, rum and Coke, whiskey on the rocks, and gin and tonic. They should have poured 44.3 ml (1.5 ounces) for each of the drinks. After pouring all of the drinks, participants were asked to estimate how much they thought they had poured, and the volume actually poured was measured.

After a distraction task, the participants were shown the tumbler and the highball glass in a rotated order and asked to estimate the total capacity of each glass.

Analysis of variance indicated that the type of drink and the interactions between the type of drink and the independent variables and covariates were not significant (P > 0.10) for either the actual or the perceived volumes poured. Because none of the covariates had a main effect on the amount of liquor poured (P > 0.10), the data were pooled.

Attention to pouring alcohol

Of 95 Philadelphia bartenders (62% men) who were approached on a Sunday or Monday evening and offered $4.00 (£2.30, €3.40) to take part in a study on "alcohol and other consumer behaviour related issues," 86 agreed to participate (62% men). They had an average of 6.3 years of bartending experience.

A 2×2 between subjects design manipulated glass shape (short and wide v tall and slender) and the amount of attention (low v high) allocated to the pouring task. Each bartender was asked to pour the established standard amount of alcohol (44.3 ml) using 1500 ml bottles and glasses as in the study of college students.

Bartenders in the low attention condition were simply asked to pour the amount of rum in a rum and Coke, the amount of gin in a gin and tonic, the amount of vodka in a vodka tonic, and the amount of whiskey in a whiskey on the rocks. The order in which they were asked to pour the drinks was randomised. Bartenders in the high attention condition were asked to pour the same four drinks, but the experimenter encouraged them to "please take your time" before they poured each drink. After this, they were asked to indicate on a nine point scale whether they agreed with the statement that they "had paid close attention to how much they poured."

A repeated measures analysis of variance indicated that there were no main effects or interactions across the types of drinks or the order poured, so the data were pooled.

Results

Practice in pouring alcohol

Overall, the college students believed the tall, slender 355 ml glasses held significantly more than the short, wide 355 ml glasses (mean 346.7 v 329.9 ml; P < 0.05, table 1), and this visual estimation bias corresponded to an opposite bias when they were pouring. They poured 30% more into short, wide glasses than tall, slender glasses (59.1 v 45.5 ml; P < 0.01). The general tendency to pour more than a 44.3 ml shot was greatest with short, wide glasses, but participants who poured into these glasses believed they poured less than those who poured into the tall, narrow glasses (44.6 v 46.1 ml; P < 0.01).

The shape of glasses continued to influence those who had done 10 practice pours only moments earlier (42.2 v 60.9 ml; P < 0.01). Although practice reduced the tendency to overpour into tall glasses (48.9 v 42.2 ml; P < 0.05), it did not do so for the short, wide glasses (60.9 v 57.3 ml; P > 0.10).

Attention to pouring alcohol

Bartenders in the high attention condition took about twice as long to pour each drink as those in the low attention condition (mean 3.7 v 1.9 seconds; P < 0.001), and they agreed more strongly with the statement that they "had paid close attention to how much they poured" (mean score 2.0 v 7.1 (maximum 9); P < 0.01).

Table 2 Shape of glass and amount of alcohol poured by bartender under low attention and high attention conditions

| | Mean (SD) amount poured (ml) | | | | | | Significance | | | | | |
| | Tall, slender glass | | | Short, wide glass | | | Glass shape | | Attention | | Attention × glass shape | |
Drink	Low attention	High attention	Average	Low attention	High attention	Average	F value	P value	F value	P value	F value	P value
All drinks	47.9 (2.6)	44.9 (2.4)	46.4	59.4 (10.8)	49.7 (3.7)	54.6	31.91	<0.01	20.19	<0.01	9.16	<0.01
Rum	48.1 (2.6)	45.2 (3.6)	46.7	60.2 (11.3)	48.4 (4.9)	54.3	24.43	<0.01	22.44	<0.01	11.86	<0.01
Vodka	47.5 (2.9)	44.6 (3.0)	46.0	59.4 (9.9)	49.8 (4.9)	54.6	37.38	<0.01	20.77	<0.01	10.23	<0.01
Whiskey	46.9 (3.8)	44.7 (3.0)	45.8	58.7 (11.2)	50.8 (5.0)	54.8	29.81	<0.01	10.58	<0.01	5.88	<0.05
Gin	49.1 (4.7)	45.0 (4.2)	47.1	59.5 (13.8)	49.8 (6.6)	54.7	15.15	<0.01	12.82	<0.01	3.68	0.08

More experienced bartenders poured an average of 10.3% less alcohol than less experienced bartenders (48.2 v 53.1 ml; $P<0.05$).

Despite an average of 6.3 years of experience, bartenders poured 20.5% more into short, wide glasses than tall, slender glasses (55.5 v 46.1 ml; $P<0.001$) (figure). The normative bias was to overpour into short, wide glasses rather than to underpour into tall, slender glasses (table 2).

Bartenders who paid less attention while pouring poured more into the short, wide glasses than into the tall, slender glasses (59.4 v 47.9 ml; $P<0.01$). If they paid careful attention while pouring, the effect was reduced (49.7 v 44.9 ml; $P<0.01$) but not eliminated.

Discussion

Although people believe they have poured more into a tall, slender glass, even professional bartenders unknowingly pour 20-30% more alcohol into short, wide glasses than into tall, slender ones. This bias is only slightly reduced by practice, concentration, or experience. Although our studies focused on pouring, both laboratory and field studies show that what is typically poured is typically drunk,[11] especially when served by a bartender.[12]

Implications for controlling alcohol consumption

This 20-30% overpouring that glass shapes can encourage needs to be accounted for in analyses of self reports of "standard" drinks. In a large epidemiological study, alcohol consumption per glass could be under-reported by as much a quarter. To account for or to correct such biases, additional questions should be added to surveys that use self reports. People drinking spirits should be asked the type or shape of glasses they typically drink from (short and wide or tall and slender), and they should be asked whether they pour freehand or with the help of a measurement aid (such as a shot glass). This information can then be used to adjust reported alcohol consumption to better reflect the actual level of consumption.

A wide range of people would like better control of alcohol consumption because of the negative consequences related to overconsumption. Those in the hospitality industry want to decrease costs (via serving size) without decreasing satisfaction. Those in public policy want to increase safety. Those dealing with alcohol counselling want to increase responsible drinking and decrease alcohol misuse.

What is already known on this topic

People pour 20-30% more into short, wide glasses than into tall, slender glasses, but they wrongly believe that tall glasses hold more

What this study adds

Bartenders poured 20% more into tumblers than into highball glasses of the same volume

Studies using self reports of "standard drinks" should ask questions about the shape of the glass

Two easy solutions to overpouring are to use or request tall, slender glasses or to use glasses on which the alcohol level is marked

If short tumblers lead even bartenders to pour more alcohol than tall highball glasses, the way to better control alcohol consumption is to use tall glasses or to use glasses with the alcohol level marked on them—and to realise that, when alcoholic drinks are served in a short wide glass, two drinks are actually equal to two and a half.

BM and KvI both contributed to design, data collection, analysis, and writing of the paper, and are guarantors.

Funding: None.

Competing interests: None declared.

Ethical approval: Standard consent forms were signed and were sufficient for institutional approval.

1 Lemmens PH. The alcohol content of self-report and "standard" drinks. *Addiction* 1994;89:593-601.
2 Carruthers SJ, Binns CW. The standard drink and alcohol consumption. *Drug Alcohol Rev* 1992;11:363-70.
3 Lemmens PH, Tan ES, Knibbe RA. Measuring quantity and frequency of drinking in a general population survey: a comparison of 5 indices. *J Stud Alcohol* 1992;53:476-86.
4 Turner C. How much alcohol is in a "standard drink"? *Br J Addiction* 1990;85:1171-5.
5 Miller WR, Heather N, Hall W. Calculating standard drink units: international comparisons. *Br J Addiction* 1991;86:43-7.
6 Stockwell T, Stirling L. Estimating alcohol content of drinks: common errors in applying the unit system. *BMJ* 1989;198:571-2.
7 Wansink B, Van Ittersum K. Bottoms up! The influence of elongation on pouring and consumption volume. *J Consumer Res* 2003;30:455-63.
8 Piaget J. *The mechanisms of perception*. London: Routledge & Kegan Paul, 1969.
9 Raghubir P, Krishna A. Vital dimensions in volume perception: can the eye fool the stomach? *J Marketing Res* 1999;36:313-26.
10 Wansink B, Cheney MM. Super bowls: serving bowl size and food consumption. *JAMA* 2005;293:1727-8.
11 Wansink B. Can package size accelerate usage volume? *J Marketing* 1996;60(3):1-16.
12 Stockwell T, Blaze-Temple D, Walker C. The effect of "standard drink" labeling on the ability of drinkers to pour a "standard drink." *Aust J Public Health* 1991;15:56-63.

Just say "No"

I was amused by a consent quandary that arose as I was about to anaesthetise a patient. Her wristband with the exaltation "Say No To Drugs" was at odds with my objective to administer potent substances intravenously through the device immediately below it. After a brief discussion, agreement was reached that I should proceed.

James Craggs *consultant anaesthetist, Lincoln County Hospital, Lincoln* (james.craggs@ulh.nhs.uk)

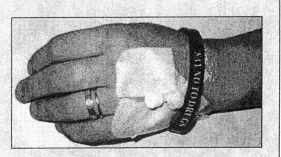

"Linking Superiority Bias in the Interpersonal and Intergroup Domains," Matthew J. Hornsey, *The Journal of Social Psychology*, 143(4):479–491, 2003.[1]

1. Who were the subjects in the two studies? What kind of a sample is this? What problems might this cause (if any)?
2. Cronbach's alpha is mentioned on p. 483. Cronbach's alpha is a slight adjustment to the average correlation of different questions on the survey. If it is 1, it means that all of the questions were perfectly correlated with what you were trying to measure, so high values say that all (or most) of the questions on the survey were consistent in their measurements.
3. On p. 483 under Analyses, they say that a mean of 1.40 shows that participants viewed themselves in a more positive light. How does this relate to the Likert-type scale of 1–9 described on p. 482 under Measures?
4. In the same section, they give a 95% confidence interval for the mean of (1.23, 1.57). Does this tell you what percent had a positive score? How would you find this percent?
5. Repeat this question for the confidence interval (0.97, 1.35) for attitudes about Australians in general.
6. For hypothesis 3, the article says "$r(82) = .47$, $p=.000$." What does the 82 mean in this context? Where did p come from? What does the value of p tell us? Is 0.47 a particularly large value?
7. Why was Study 2 conducted? How did Study 2 differ from Study 1?
8. On p. 485 under Measures and Scales, they talk about principal component analysis for PSES. Principal components is a way to combine a number of responses into one (or a few) "indices" that contain most of the original information. This was also done for CSES, so we can consider each of these tests to produce a single number.
9. On p. 486 under Results, they say, "The 95% confidence intervals ... did not span zero." Why is this important? What other information is in this statement?
10. Comment on the order effect where $M=0.91$ in one order and $M=1.06$ in the other order. What test would you use for these? How would you compute a confidence interval?
11. Table 2 gives correlations among the four measures. All of these values are positive. Is that to be expected? The three largest values are significant. Would it always be the case that if some of the values are significant, it would be the larger values? When might this not be the case?

---

[1] © 2003, Heldref Publications. Used by permission.

Linking Superiority Bias in the Interpersonal and Intergroup Domains

MATTHEW J. HORNSEY
School of Psychology
University of Queensland
St. Lucia, Australia

ABSTRACT. The author conducted 2 studies to explore the link between superiority bias in the interpersonal and intergroup domains. Australian university students evaluated the extent to which various personality traits were more or less applicable to themselves than to other Australian university students in general. They then evaluated the extent to which the same traits were more or less applicable to Australians than to people from other countries in general. As expected, the more participants evaluated themselves as superior to other university students, the more they evaluated Australians as a whole as superior to people from other countries. This link between interpersonal and intergroup superiority biases explained 22.1% of variance in Study 1 and 33.6% of variance in Study 2. The author interprets the results of the 2 studies as support for fundamental principles of social identity theory: (a) that self-concept consists of not only one's personal self but also the social groups to which one belongs and (b) that people are motivated to view both levels of self in a relatively positive fashion.

Key words: intergroup bias, social identity theory, superiority bias

WHEN EVALUATING THE SELF relative to others, people demonstrate a wide range of self-serving biases. People consider themselves to be less likely than others to experience negative events in the future (*unrealistic optimism*; Perloff & Fetzer, 1986; Weinstein, 1980). People believe themselves to be less influenced than others by negative media messages (*third-person effect*; e.g., Davison, 1983; Gunther, 1995; McLeod, Eveland, & Nathanson, 1997) but more influenced by prosocial messages (e.g., David & Johnson, 1998; Duck, Hogg, & Terry, 1999). Finally, people attribute positive personality traits more to themselves than to other people in general (*illusory superiority*; e.g., Alicke, 1985; Brown, 1986).

This research was funded by a postdoctoral fellowship from the Australian Research Council. Thanks to Jolanda Jetten and Sabine Otten for their helpful comments on an earlier version of this article.
Address correspondence to Matthew J. Hornsey, School of Psychology, University of Queensland, St. Lucia 4072, Queensland, Australia; m.hornsey@psy.uq.edu.au (e-mail).

Illusory superiority exists on dimensions as diverse as honesty, attractiveness, persistence, independence, and sincerity, and the bias gets larger as the trait is seen to be more desirable.

Recently, a number of researchers have proposed that these phenomena may be artifacts of cognitive processes, that superiority biases are driven by the cognitive strain involved with making self–other comparisons (e.g., Kruger, 1999). Although these cognitive explanations have merit, strong circumstantial evidence has supported the notion that biases have at least some motivational basis (Hoorens, 1993; Kunda, 1990; Kunda & Sanitioso, 1989). The motivational view argues that superiority bias is not necessarily a conscious strategy that people use to maintain a positive self-concept. Consistent with this view, on dimensions that cannot be objectively evaluated (e.g., personality dimensions), biases have almost always occurred in a self-flattering direction (see Hoorens; Kunda, for reviews). Furthermore, illusory superiority has been positively correlated with self-esteem and has not been displayed at all among individuals who are depressed (Brown, 1986). On the basis of such research, illusory superiority can be viewed as a healthy and adaptive way to achieve self-validation and to maintain mental health (Taylor & Brown, 1988).

The desire to see oneself in a relatively positive light applies also to more extended definitions of self. Much as people have rated themselves favorably relative to others, they also have evaluated their friends relatively favorably (Brown, 1986). It is assumed that part of the reason for this is that people view their friends as part of their extended selves. Another study that has alluded to the link between self-concept and people's perception of important others was conducted by Heine and Lehman (1997), who compared self-serving biases and family-serving biases among Canadians. To measure self-serving bias, participants estimated the percentage of the population that was better than they with respect to 10 traits. To measure family-serving bias, participants were first asked to write down the name of the member of their family to whom they felt closest. Family-serving bias was then calculated by estimating the percentage of the population that was better than the family member on the 10 traits. Heine and Lehman found that for European Canadians, there was a strong positive correlation ($rs = .35-.53$) between self-serving and family-serving biases. Those authors interpreted this finding as evidence that Canadians were "basking in the reflected glory of their family members. Saying, for example, that their mom is better than most others on a variety of traits apparently makes European Canadians feel better about themselves" (p. 1279).

The notion that there might be a link between evaluations of the self and evaluations of significant others is not new to researchers of group identity. Proponents of social identity theory (SIT; Tajfel & Turner, 1979; Turner, 1999) have proposed that people draw identity out of the groups to which they belong. Crucially, the way in which the group is evaluated affects one's self-esteem. Assuming that people have a fundamental need to see themselves in a relatively positive light, SIT proponents argue that self-enhancement is best attained by strategies that achieve or maintain a sense of in-group superiority.

Some researchers and theorists consider the fundamental need to achieve positive self-regard as the underlying reason that people evaluate groups to which they belong (the in-group) more positively than they evaluate other people's groups (out-groups). As observed by Sumner (1906), "one's group is the center of everything, and all others are scaled and rated with reference to it. . . . Each group nourishes its own pride and vanity, boasts itself superior, exalts its own divinities, and looks with contempt on outsiders" (p. 13). Proponents of SIT are not quite as dogmatic as Sumner on this matter; they acknowledge that the propensity to show in-group bias is often affected by reality constraints, perceptions of threat, and sociostructural variables such as status, power, legitimacy, and permeability of group boundaries. However, as a general principle, perceptions of one's group are typically skewed in the positive direction. Such a bias is pervasive both in real life and in the laboratory and has even emerged when the basis for categorization was random or trivial (e.g., Billig & Tajfel, 1973; Diehl, 1990).

The similarities between interpersonal forms of superiority bias and the group-oriented bias described above are striking. In both cases, some researchers and theorists have assumed that people are motivated, for self-esteem reasons, to have a positive sense of their relative worth. The key difference is in relation to what level of self is being considered. In the case of interpersonal superiority bias, the *self* is defined in a traditional sense, as a personal and unique accumulation of generalizations, beliefs, and experiences. In the case of in-group bias, the self is defined more in terms of one's social identity: the norms, attitudes, and behaviors prescribed by one's social group membership and the value associated with them.

The aim of the present study was to draw parallels, both conceptually and empirically, between superiority bias in the interpersonal and intergroup domains. Despite the obvious similarities between the two types of superiority bias, they tend to occupy separate research traditions, and they rarely refer to one another. Bridges between these two research traditions may help provide a fresh perspective on the study of self-serving biases.

STUDY 1

In Study 1, participants evaluated the extent to which various personality traits are more or less applicable to themselves than to "other Australian university students in general." Participants then evaluated the extent to which the same traits are more or less applicable to Australians than to people from "other countries in general."

On the basis of previous research, I expected that participants would show superiority bias on both the interpersonal level (Hypothesis 1) and the intergroup level (Hypothesis 2). Because both forms of bias are believed to be at least partly driven by a need for a positive self-concept, it is reasonable to expect that the degrees of bias shown at the interpersonal and intergroup levels would be linked. Specifically, I expected that there would be a significant positive correlation between the degree to which participants demonstrated interpersonal superiority bias and the degree to which they demonstrated intergroup superiority bias (Hypothesis 3).

Method

Participants

Participants were 85 Australian psychology students (22 men, 63 women; $M = 18.80$ years, $SD = 2.72$ years). They participated in return for course credit.

Measures

Interpersonal superiority bias was measured with a technique of Alicke (1985) and Brown (1986). Participants used a 9-point Likert-type scale to evaluate the extent to which 18 personality dimensions applied to themselves more or less than to other Australian university students in general (1 = *much less applicable to me than to others*, 9 = *much more applicable to me than to others*). Australian university students were chosen as the comparison group rather than Australians because there are some traits (e.g., intelligence) in which university students might differ from the rest of the population. By using a comparison group that had the same level of education as the participants, researchers can more readily attribute any biases that are found to motivational variables rather than real differences. The personality traits were selected from a list of desirable and undesirable traits compiled by Anderson (1968). The positive traits were *trustworthy, honest, broad-minded, happy, well-balanced, nice, intelligent, good-humored,* and *friendly*. The negative traits were *submissive, cruel, short-tempered, superficial, egocentric, rude, possessive, touchy,* and *withdrawn*. To guard against response bias, I presented the positive and negative traits alternately.

After completion of this scale, participants evaluated the extent to which the same traits applied more or less to Australians than to people in general from other countries (1 = *much less applicable to Australians than to others*, 9 = *much more applicable to Australians than to others*). I interpreted the responses on this set of items as indices of intergroup bias. After completion of this questionnaire, participants were debriefed.

Results and Discussion

Scales

I reversed scores on the negative traits so that for all items, larger values represented more positive evaluations relative to others. In preliminary analyses, I found no difference in the pattern of results for the negative and positive traits.

Consequently, the main analyses were performed with the negative and positive traits combined. Ratings on the 18 personality traits formed a stable, strong scale both for ratings of self relative to other university students (Cronbach's $\alpha = .82$) and for ratings of Australians relative to people from other countries (Cronbach's $\alpha = .88$). In each case, the items were recoded so that positive scores represented perceptions of superiority relative to others, negative scores represented perceptions of inferiority relative to others, and a score of zero represented no discrepancy between perceptions of self and others.

Analyses

Consistent with Hypothesis 1, participants viewed themselves in a more positive light relative to other university students ($M = 1.40$, $SD = 0.77$). Inspection of the 95% confidence interval (1.23, 1.57) showed that this score was clearly different from zero. Consistent with Hypothesis 2, participants also showed a clear tendency to view Australians in a more positive light than people from other countries ($M = 1.16$, $SD = 0.85$). Again, the 95% confidence interval (0.97, 1.35) did not span zero, indicating a significant level of bias.

Hypothesis 3 stated that there would be a significant, positive correlation between bias in the interpersonal domain and bias in the intergroup domain. This hypothesis was supported, $r(82) = .47$, $p = .000$, $r^2 = .221$. Consistent with expectations, the more participants evaluated themselves to be superior relative to other university students in general, the more they evaluated Australians to be superior relative to people from other countries.

In summary, Study 1 suggested links between the interpersonal and intergroup biases among Australian university students. There is no practical reason that people who consider themselves to be superior to other individuals should also consider their compatriots to be superior to citizens of other countries. On the contrary, one may logically expect the opposite effect. If certain Australians think that they are superior to other Australians, then it would be reasonable to assume that they might have a relatively poor image of Australians relative to people from other cultures. The positive relationship between interpersonal superiority and intergroup superiority suggests that both biases are underpinned by similar psychological motives: presumably, the need to achieve or maintain a positive self-concept.

STUDY 2

The results of Study 1 demonstrated a significant correlation between interpersonal superiority bias and intergroup superiority bias. One aim of Study 2 was to test the robustness of this effect (across time and different samples) by means of a replication. Designed partly for the standard empirical advantages of replication, Study 2 was designed also to reexamine the relationship between the two forms of superiority bias after the controlling for possible order effects. In Study 1, participants always filled out the questionnaire examining interpersonal comparisons before completing the intergroup comparisons. In Study 2, the order of presentation of the items was counterbalanced.

Study 2 also investigated how interpersonal and intergroup superiority biases correlate with self-esteem. As described earlier, superiority biases have often been interpreted as a healthy and adaptive way to maintain a positive self-concept (Taylor & Brown, 1988). Consistent with this interpretation, superiority bias in the interpersonal domain has been positively correlated with self-esteem (Brown, 1986). The link between self-esteem and intergroup bias tells a similar story. A recent meta-analysis (Aberson, Healy, & Romero, 2000) has indicated that generally, people with high self-esteem tend to show more in-group bias.

However, this latter link has been generally weak (Rubin & Hewstone, 1998), leading some researchers and theorists to question the methods used to measure self-esteem in intergroup contexts. One criticism of previous work on the link between self-esteem and in-group bias has been that there is a discrepancy between the levels of abstraction at which self-esteem and in-group bias have been measured (Breckler & Greenwald, 1986; Long & Spears, 1997; Luhtanen & Crocker, 1992). Specifically, proponents of SIT argue that maintaining a positive view of one's group enhances one's *social* identity, the part of one's self-concept that is described and defined by one's social group memberships. So, on the surface, personal self-esteem is an inappropriate predictor of in-group bias, because it is defined at the individual level. Consequently, a number of theorists have argued that self-esteem should be measured at the collective level, the most popular version of which is Luhtanen and Crocker's Collective Self-Esteem Scale (CSES).

Despite the observation that the CSES appears to be a more theoretically appropriate predictor of in-group bias, the evidence for its utility is mixed. Indeed, in their meta-analysis, Aberson et al. (2000) found that personal self-esteem was a stronger predictor of in-group bias than was the CSES. It might, however, be a little premature to disregard the distinction between personal self-esteem and collective self-esteem. The disappointing relationship between collective self-esteem and in-group bias may reflect the abstract reference of the CSES items to people's social groups. To clarify the meaning of *social groups*, the instructions for the CSES invite people to consider their "gender, race, religion, nationality, ethnicity, and socioeconomic class" (Luhtanen & Crocker, 1992). As pointed out by Rubin and Hewstone (1998), this very global measure of self-esteem is not just cognitively complex but also ill equipped to predict social comparisons made with regard to a specific group. They recommended that the items be adapted to refer to the specific social identity in question.

Rubin and Hewstone's (1998) theoretical review identified only three investigations of the relationship between bias and self-esteem in which researchers adapted the CSES to relate to a specific identity. The results of these studies have been contradictory. Maass, Ceccarelli, and Rudin (1996) found that collective self-esteem correlated positively with linguistic bias among hunters and environmentalists (Experiment 1), but they found no relationship between collective self-esteem and linguistic bias among northern and southern Italians (Experiment 2). Similarly, Long, Spears, and Manstead (1994) found no clear relationship between the collective self-esteem of Dutch participants and in-group bias in relation to people from other countries. Thus, although it makes good theoretical sense to measure self-esteem and bias at the same level of abstraction, evidence is conflicting about whether collective self-esteem and intergroup bias are linked.

In summary, the aims of Study 2 were (a) to replicate the results of Study 1 after controlling for order effects and (b) to test the relationship between self-esteem and superiority bias. As in Study 1, participants in Study 2 evaluated themselves relative to other university students and evaluated Australians in relation to people from other countries. On the basis of Study 1, I expected that participants would demonstrate superiority biases at both the interpersonal and intergroup levels. Furthermore, it was expected that the more participants showed bias at the interpersonal level, the more they would show bias at the intergroup level.

On the basis of previous research (e.g., Aberson et al., 2000; Brown, 1986), I expected bias to be positively related to self-esteem, so that people would show more bias the higher their self-esteem. Furthermore, on the basis of theoretical arguments made by Long and Spears (1997) and Rubin and Hewstone (1998), I expected that self-esteem and bias would correlate more strongly when I measured them at the same level of abstraction. Specifically, I expected that interpersonal bias would correlate more highly with personal self-esteem, whereas in-group bias would correlate more highly with collective self-esteem.

Method

Participants

Participants were 40 Australian psychology students (16 men, 24 women; $M = 19.70$ years, $SD = 4.09$ years). They participated in the study in return for course credit.

Measures and Scales

I first measured self-esteem and then measured superiority bias. Self-esteem was assessed with two measures: Rosenberg's (1965) 10-item Private Self-Esteem Scale (PSES) and Luhtanen and Crocker's (1992) CSES. Half of the participants completed the PSES first, and the other half completed the CSES first. On the basis of a scree test, principal components analysis on the PSES revealed a single factor (eigenvalue = 4.74). So, these items were combined into a single measure of personal self-esteem (Cronbach's $\alpha = .86$). As recommended by Maass et al. (1996), only those subscales of the CSES Private Collective Self-Esteem and Importance to Identity that were of particular relevance to analyses of group identity were measured. Furthermore, as recommended by Long and Spears (1997) and Rubin and Hewstone (1998), these eight items were adapted to relate specifically to the group identity in question. Thus, instead of referring to participants' social groups in general, the questions were tailored to apply to the identity of being an Australian. When considered together, the two subscales formed a single, stable factor (eigenvalue = 4.66) and so were combined into a scale of collective self-esteem (Cronbach's α = .84).

Superiority bias was measured by the same method described in Study 1. As in Study 1, the ratings on the personality dimensions were combined into reliable scales of superiority for the interpersonal (Cronbach's α = .83) and intergroup domains (Cronbach's α = .86). To neutralize any order effects, half the participants completed the interpersonal measures first, and the other half completed the intergroup measures first.

Results and Discussion

As Table 1 indicates, participants viewed themselves in a more positive light relative to other university students ($M = 1.17$, $SD = 0.83$) and viewed Australians as superior to people from other countries ($M = 0.97$, $SD = 0.95$). The 95% confidence intervals for interpersonal bias (0.90, 1.44) and intergroup bias (0.66, 1.29) did not span zero, indicating significant levels of bias. The amount of interpersonal bias did not differ according to whether it was measured before ($M = 1.13$, $SD = 0.93$) or after intergroup comparisons ($M = 1.23$, $SD = 0.70$), $t(37) = 0.39$, $p = .70$, $\eta^2 = .00$. Similarly, the amount of intergroup bias did not differ according to whether it was measured before ($M = 0.91$, $SD = 1.03$) or after interpersonal comparisons ($M = 1.06$, $SD = 0.85$), $t(36) = 0.45$, $p = .66$, $\eta^2 = .01$.

Table 2 summarizes the correlations among the measures of bias and self-esteem. As in Study 1, there was a significant, positive correlation between bias in the interpersonal domain and bias in the intergroup domain, $r(37) = .58$, $p = .000$, $r^2 = .336$. As in Study 1, the more participants evaluated themselves to be superior relative to other Australian university students, the more they evaluated Australians to be superior relative to people from other countries. This result suggested that the relationship demonstrated in Study 1 was both large and reasonably reliable. Furthermore, the relationship was not affected by the order in which interpersonal and intergroup comparisons are made.

A secondary aim of Study 2 was to examine how superiority biases covary with self-esteem. As expected, interpersonal superiority bias significantly correlated with PSES, $r(39) = .47$, $p = .002$, $r^2 = .221$, so that the higher was their self-esteem, the more they showed bias. The relationship between PSES and intergroup superiority bias was not as impressive, $r(38) = .23$, $p = .17$, $r^2 = .053$. However, when measured at the collective level of abstraction (CSES), the relationship between self-esteem and in-group bias was significant, $r(38) = .46$, $p = .004$, $r^2 = .212$. Unfortunately, the relatively small sample size did not afford

TABLE 1. Means and Standard Deviations for Participants' Scores in Study 2

Statistic	IP bias	IG bias	PSES	CSES
M	1.17	0.97	7.00	6.56
SD	0.83	0.95	1.08	1.65

Note. IP = interpersonal. IG = intergroup. PSES = Private Self-Esteem Scale (Rosenberg, 1965). CSES = Collective Self-Esteem Scale (Luhtanen & Crocker, 1992).

TABLE 2. Correlations of Participants' Scores in Study 2

Measure	1	2	3	4
1. IP bias	—			
2. IG bias	.58**	—		
3. PSES	.47*	.23	—	
4. CSES	.32	.46*	.27	—

Note. IP = interpersonal. IG = intergroup. PSES = Private Self-Esteem Scale (Rosenberg, 1965). CSES = Collective Self-Esteem Scale (Luhtanen & Crocker, 1992).
*$p < .01$. **$p < .001$.

enough power to allow a definitive test of the difference between these correlations. Thus, although CSES explained four times as much variance in intergroup bias as did PSES, a test of the difference between these correlations revealed no significant effect, $t(37) = 1.30$, $p = .20$. Despite this lack of statistical closure, the pattern of correlations is broadly supportive of the perspective that when predicting intergroup bias, it is most appropriate to measure self-esteem at the collective level of abstraction—a notion that has had mixed support in the literature (Rubin & Hewstone, 1998).

In summary, the correlational data suggested another link between the interpersonal and intergroup forms of superiority bias; not only were they strongly correlated, but they each covaried in a predictable way with self-esteem. However, there is preliminary evidence that interpersonal and intergroup biases covary with self-esteem, but only when self-esteem is measured at an appropriate level of abstraction. Specifically, interpersonal bias was best predicted by personal self-esteem, and intergroup bias was best predicted by collective self-esteem.

GENERAL DISCUSSION

The aim of the present study was to draw conceptual and empirical parallels between interpersonal superiority bias and intergroup bias. Conceptually, I have argued that the two forms of bias share a similar motive: a need for a positive self-concept. Those who research the superiority bias have explicitly argued that self-favoring comparisons are made to create and maintain a positive view of oneself. Similarly, social identity theorists claim that the value that people attach to their group memberships has direct implications for their self-concept. Consequently, people are motivated to think of their groups as being more positive than other groups, because such a bias helps create and maintain a positive view of oneself.

If it is true that the two phenomena share a similar underlying motive—to have a positive self-concept—one can reasonably expect that the biases at the interpersonal and intergroup levels would be empirically linked. Among our sample of Australian university students, this proved to be the case. In both of the present studies, the more people viewed themselves as superior to other Australian university students, the more they considered Australians as superior to citizens of other countries in general. This relationship was not just consistent, it was also relatively large. In Study 1, interpersonal bias accounted for 22.1% of variance in intergroup bias; in Study 2, it accounted for 33.6%.

Care should be taken, however, not to make simplistic assumptions on the basis of this relationship. First, it would be absurd to suggest that all types of in-group bias are simply alternative forms of self-favoring bias that are based on the same psychological processes as, for example, illusory superiority. One of the original missions of SIT (Tajfel & Turner, 1979; Turner, 1999) was to enable one to take into account the emergent processes of group life without reference to individualistic explanations. Consistent with this mission, it has been demonstrated conclusively that in-group bias is affected by a number of purely group-related variables, such as identity salience (e.g., Turner, Hogg, Oakes, Reicher, & Wetherell, 1987), interdependence (e.g., Bourhis, Turner, & Gagnon, 1997), status (e.g., Tajfel, 1978; Tajfel & Turner, 1979), power (e.g., Ng, 1980; Sachdev & Bourhis, 1991), and identity threat (Branscombe, Ellemers, Spears, & Doosje, 1999).

Second, in contrast to real-world instances of intergroup competition, the present study's definition of bias lacked reference to a specific out-group. Research on self-favoring biases shows that the more specific is the comparison other, the smaller is the bias (Alicke, Klotz, Breitenbecher, Yurak, & Vredenburg, 1995). Given the close parallels that have been demonstrated between interpersonal and intergroup bias, it seems likely that the magnitude of in-group bias would be affected also by the specificity of the comparison's other. Bias between two specific groups may increase or decrease depending on the history of the relationship and the sociostructural context. Given this extra complexity, it is unlikely that bias between two specific groups could be as cleanly predicted by constructs such as interpersonal bias and self-esteem as it was here. This complexity may help explain the discrepancy between the results reported here and those reported by Long et al. (1994) and by Maass et al. (1996; Experiment 2), where no significant correlation was found between collective self-esteem and in-group bias.

Finally, I hope that the present study will not be interpreted as evidence that in-group bias is driven entirely by an individualistic drive for personal self-esteem. Rather, the evidence suggests that positive self-regard means different things in the interpersonal and intergroup contexts. Whereas personal self-esteem correlated most strongly with perceptions of interpersonal superiority, perceptions of intergroup superiority correlated most strongly with feelings about one's social group. These findings of Study 2 underscored the need for researchers to use measures of collective self-esteem when examining motives for ethnocentrism and in-group bias (see also Long & Spears, 1997; Rubin & Hewstone, 1998).

If it is true that interpersonal and intergroup superiority biases are associated with self-esteem at different levels of abstraction, then perhaps people could protect their self-concept by strategically shifting from one level of identity to another. For example, if an individual's personal self-esteem is threatened (e.g., because he or she is inadequate in a task), it might be possible for the person to maintain a positive self-concept by perceiving positive differences between his or her group and other groups. Similarly, if a person's collective self-esteem is threatened (e.g., because his or her group is low in status or is inadequate), it might be possible for the person to maintain a positive self-concept by perceiving positive differences between the self and other individuals within the group. An experimental paradigm—in which the positivity of interpersonal and intergroup identities is manipulated—would be a logical extension of the correlational data reported here.

In summary, the current data helped make clear certain theoretical similarities between two streams of literature that have had little comparison in the past. Superiority biases in the interpersonal and intergroup domains are not the same phenomena, but they have been empirically linked. This finding supports very simple, classical SIT notions: (a) that self-concept reflects not just one's personal self but also the social groups that one belongs to and (b) that people are motivated to view all levels of self in a relatively positive light.

REFERENCES

Aberson, C. L., Healy, M., & Romero, V. (2000). Ingroup bias and self-esteem: A meta-analysis. *Personality and Social Psychology Review, 4*, 157–173.

Alicke, M. D. (1985). Global self-evaluation as determined by the desirability and controllability of trait adjectives. *Journal of Personality and Social Psychology, 49*, 1621–1630.

Alicke, M. D., Klotz, M. L., Breitenbecher, D. L., Yurak, T. J., & Vredenburg, D. S. (1995). Personal contact, individuation, and the better-than-average effect. *Journal of Personality and Social Psychology, 68*, 804–825.

Anderson, N. H. (1968). Likableness ratings of 555 personality trait words. *Journal of Personality and Social Psychology, 9*, 272–279.

Billig, M., & Tajfel, H. (1973). Social categorization and similarity in intergroup behavior. *European Journal of Social Psychology, 3,* 27–52.

Bourhis, R. Y., Turner, J. C., & Gagnon, A. (1997). Interdependence, social identity and discrimination. In R. Spears, P. Oakes, N. Ellemers, & S. Haslam (Eds.), *The social psychology of stereotyping and group life* (pp. 273–295). Oxford, England: Blackwell.

Branscombe, N., Ellemers, N., Spears, R., & Doosje, B. (1999). The context and content of social identity threat. In N. Ellemers, R. Spears, & B. Doosje (Eds.), *Social identity: Context, commitment, content* (pp. 35–58). Oxford, England: Blackwell.

Breckler, S., & Greenwald, A. G. (1986). Motivational facets of the self. In R. L. Sorrentino & E. T. Higgins (Eds.), *Handbook of motivation and cognition* (pp. 145–164). New York: Guilford.

Brown, J. D. (1986). Evaluations of self and others: Self-enhancement biases in social judgments. *Social Cognition, 4,* 353–376.

David, P., & Johnson, M. A. (1998). The role of self in third-person effects about body image. *Journal of Communication, 48,* 37–58.

Davison, W. P. (1983). The third-person effect in communication. *Public Opinion Quarterly, 47,* 1–15.

Diehl, M. (1990). The minimal group paradigm: Theoretical explanations and empirical findings. *European Review of Social Psychology, 1,* 263–292.

Duck, J. M., Hogg, M. A., & Terry, D. J. (1999). Social identity and perceptions of media persuasion: Are we always less influenced than others? *Journal of Applied Social Psychology, 29,* 1879–1899.

Gunther, A. C. (1995). Overrating the X-rating: The third-person perception and support for censorship of pornography. *Journal of Communication, 45,* 27–38.

Heine, S. J., & Lehman, D. R. (1997). The cultural construction of self-enhancement: An examination of group-serving biases. *Journal of Personality and Social Psychology, 72,* 1268–1283.

Hoorens, V. (1993). Self-enhancement and superiority biases in social comparison. *European Review of Social Psychology, 4,* 113–139.

Kruger, J. (1999). Lake Wobegon be gone! The "below-average effect" and the egocentric nature of comparative ability judgments. *Journal of Personality and Social Psychology, 77,* 221–232.

Kunda, Z. (1990). The case for motivated reasoning. *Psychological Bulletin, 108,* 480–498.

Kunda, Z., & Sanitioso, R. (1989). Motivated changes in the self-concept. *Journal of Experimental Social Psychology, 25,* 272–285.

Long, K., & Spears, R. (1997). The self-esteem hypothesis revisited: Differentiation and the disaffected. In R. Spears, P. Oakes, N. Ellemers, & S. Haslam (Eds.), *The social psychology of stereotyping and group life* (pp. 296–317). Oxford, England: Blackwell.

Long, K., Spears, R., & Manstead, A. S. R. (1994). The influence of personal and collective self-esteem on strategies of social differentiation. *British Journal of Social Psychology, 33,* 313–329.

Luhtanen, R., & Crocker, J. (1992). A Collective Self-Esteem Scale: Self-evaluation of one's social identity. *Personality and Social Psychology Bulletin, 18,* 302–318.

Maass, A., Ceccarelli, R., & Rudin, S. (1996). Linguistic intergroup bias: Evidence for ingroup-protective motivation. *Journal of Personality and Social Psychology, 71,* 512–526.

McLeod, D. M., Eveland, W. P., Jr., & Nathanson, A. I. (1997). Support for censorship of violent and misogynic rap lyrics: An analysis of the third-person effect. *Communication Research, 24,* 153–174.

Ng, S. H. (1980). *The social psychology of power.* New York: Academic Press.

Perloff, L. S., & Fetzer, B. K. (1986). Self-other judgments and perceived vulnerability to victimization. *Journal of Personality and Social Psychology, 50,* 502–510.

Rosenberg, M. (1965). *Society and the adolescent self-image.* Princeton, NJ: Princeton University Press.

Rubin, M., & Hewstone, M. (1998). Social identity theory's self-esteem hypothesis: A review and some suggestions for clarification. *Personality and Social Psychology Review, 2,* 40–62.

Sachdev, I., & Bourhis, R. Y. (1991). Power and status differentials in minority and majority group relations. *European Journal of Social Psychology, 21,* 1–24.

Sumner, W. G. (1906). *Folkways.* Boston, MA: Ginn.

Tajfel, H. (Ed.). (1978). *Differentiation between social groups: Studies in the social psychology of intergroup relations.* London: Academic Press.

Tajfel, H., & Turner, J. C. (1979). An integrative theory of intergroup conflict. In W. G. Austin & S. Worchel (Eds.), *The social psychology of intergroup relations* (pp. 33–47). Monterey, CA: Brooks/Cole.

Taylor, S. E., & Brown, J. D. (1988). Illusion and well-being: A social psychological perspective on mental health. *Psychological Bulletin, 103,* 193–210.

Turner, J. C. (1999). Some current issues in research on social identity and self-categorization theories. In N. Ellemers, R. Spears, & B. Doosje (Eds.), *Social identity: Context, commitment and content* (pp. 6–34). Oxford, England: Blackwell.

Turner, J. C., Hogg, M. A., Oakes, P. J., Reicher, S. D., & Wetherell, M. S. (1987). *Rediscovering the social group: A self-categorization theory.* New York: Blackwell.

Weinstein, N. D. (1980). Unrealistic optimism about future life events. *Journal of Personality and Social Psychology, 39,* 806–820.

Received April 24, 2001
Accepted November 20, 2001

"Monitoring Early Reading Development in First Grade: Word Identification Fluency Versus Nonsense Word Fluency," Fuchs, Fuchs, Compton, *Exceptional Children*, 71(1): 7–21, 2004.[1]

1. On p. 11, under Criterion Measures, the authors discuss "Split-half and test-retest reliabilities." You can read about these on the Internet. To calculate split-half reliability, you randomly divide the questions into two sets and find the correlation between scores on each half. When evaluating reliability, are you interested in significance testing? If so, what is the null hypothesis?
2. Pretend that you have administered tests to some children. Go through the steps of calculating one of the reliabilities above. Which numbers are you comparing to which? What will be your sample size?
3. In the same section, they discuss "concurrent validity" with another test. What did they calculate? What is important about this measure?
4. On p. 13, they present and discuss Table 1, summary scores for the test periods. At the bottom of the first column, they say that "the fall CBM WIF is 0.85 standard deviations below the mean." What mean are they referring to? What is the value of this mean?
5. What is noteworthy about the fall CBM WIF statistics? What does this tell you about the scores? What impact does your observation have on the interpretation of these scores?
6. In the top of the second column on p. 13, they say that "the end of year CRAB fluency score is … below … 40." Where would you find this in Table 1? What would be the *P*-value for this test? Should this be a one-sided or a two-sided test? Does it matter?
7. Table 2 on p. 14 contains measures of concurrent and predictive validity for both WIF and NWF. The last column is headed $t(148)$. What sort of statistics are in this column? What does 148 signify? Some of the values have asterisks and some do not. What does this mean?
8. The last column is for comparing correlations based on the same sample. This is not often covered in introductory statistics courses, but tests of correlation often are covered. Suppose that the NWF column had been calculated using a different set of students than the WIF column. How many df would each column have? Can you guess how many df their difference would have?
9. In this last column, was the comparison done using one-sided or two-sided tests? If one-sided, which side? How do you determine this? What difference does it make whether you do one- or two-sided tests?
10. What is the difference between the parts of Table 2 labeled Concurrent Validity and Predictive Validity?
11. The first entry under Predictive Validity is 0.63. Carefully determine exactly which sets of scores were used to produce this value.
12. Table 3 on p. 15 is discussed starting at the bottom of the first column of that page. There are four rows to this table. How were these derived? The first entry in this table is 0.60. Carefully consider which scores on which tests were used to produce this value.

---

[1] *© 2004, The Council for Exceptional Children. Reprinted with permission.*

13. The footnote to Table 3 notes that the *t* value associated with $p<.05$ is 1.976. This number sounds vaguely familiar. Why is that? Does this tell you if they are doing one-sided or two-sided tests?
14. The right half of Table 3 is labeled t(148). See the discussion of this in Table 2.
15. There are 16 values in the right half of Table 3. Should we be concerned about seeing so many statistics in one place? (Hint: Yes.) If we calculated 16 statistics at random, how many would we expect to see being significant at the 5% level?

Monitoring Early Reading Development in First Grade: Word Identification Fluency Versus Nonsense Word Fluency

LYNN S. FUCHS
DOUGLAS FUCHS
DONALD L. COMPTON
Peabody College of Vanderbilt University

ABSTRACT: *This study contrasts the validity of 2 early reading curriculum-based measurement (CBM) measures: word identification fluency and nonsense word fluency. At-risk children (n = 151) were assessed (a) on criterion reading measures in the fall and spring of first grade and (b) on the 2 CBM measures each week for 7 weeks and twice weekly for an additional 13 weeks. Concurrent and predictive validity for CBM performance level and predictive validity for CBM slopes demonstrated the superiority of word identification fluency over nonsense word fluency. Findings are discussed in terms of the measures' utility for identifying children in need of intensive instruction and for monitoring children's progress through first grade.*

Curriculum-based measurement (CBM) provides teachers with reliable, valid, and efficient indicators of academic competence with which to gauge individual student standing at one point in time or to index student progress across time (Deno, 1985). CBM is the most widely studied form of classroom assessment, with more than 150 studies in peer-reviewed journals establishing its psychometric tenability and its instructional utility. Given the strength of the existing literature, CBM is becoming a signature feature associated with effective special education (McDonnell, McLaughlin, & Morison, 1997; President's Commission, 2002).

At the same time, in reading, most CBM research has focused on the passage reading fluency task, which becomes appropriate for most students sometime during the second semester of first grade. Additional research is needed to examine the tenability of reading tasks that address an earlier phase of reading, developmentally appropriate for most children during the first half or (depending on the child) much of first grade. This study compared two CBM measures for this beginning phase of first-grade reading development. In this introduction, we provide background information on CBM, review the literature on the two measures on which we focus the present study, and clarify the purpose and importance of the present study.

BACKGROUND INFORMATION ON CBM

CBM differs from most forms of classroom assessment in several ways (Fuchs & Deno, 1991) including these two features: First, CBM is standardized so that the behaviors to be measured and the procedures for measuring those behaviors are prescribed, with documented reliability and validity. Second, CBM's focus is long-term so that the testing methods and the testing content remain constant, with equivalent weekly tests spanning much, if not all, of the school year. The reason for long-term consistency is so that progress can be monitored systematically over time.

In using the CBM passage reading fluency task, the teacher relies on established methods to identify passages of equivalent difficulty; each equivalent passage represents the material students should be comfortable reading at year's end. The teacher administers CBM by having the student read aloud a different passage for each assessment, each time for 1 min, and the weekly score is the number of words read correctly. Each assessment produces an indicator of reading competence because it requires a multifaceted performance. This performance entails, for example, a reader's skill at automatically translating letters into coherent sound representations, unitizing those sound components into recognizable wholes and automatically accessing lexical representations, processing meaningful connections within and between sentences, relating text meaning to prior information, and making inferences to supply missing information. As competent readers translate text into spoken language, they coordinate these skills in an obligatory, seemingly effortless manner (Fuchs, Fuchs, Hosp, & Jenkins, 2001).

Because the CBM passage reading fluency task reflects this complex performance, it can be used to characterize reading expertise and to track its development in the primary grades (e.g., Biemiller, 1977-1978; Fuchs & Deno, 1991). To characterize reading expertise, CBM is interpreted in a norm-referenced or criterion-referenced framework. Within a normative framework, practitioners designate reading difficulty by comparing CBM performance levels between individuals. For example, with local CBM norms, students performing below the third percentile are identified as reading disabled and entitled to intensive reading instruction. Within a criterion-referenced perspective, CBM benchmarks specify the minimum performance levels associated with future reading success. For example, schools might establish that students who fail to attain a CBM score of 75 by the end of second grade have a poor probability of achieving a "proficient" score on their state's high-stakes fourth-grade reading assessment; therefore, students scoring below this benchmark at the end of second grade are candidates for intensive reading instruction.

Across these normative- and criterion-referenced perspectives, CBM is administered in a single timeframe to identify students who require special attention. By contrast, the major purpose for CBM is to monitor the development of academic competence. Progress monitoring requires an intra-individual framework, where CBM is collected frequently (weekly or monthly); student scores are graphed; a slope is derived from the graphed scores to quantify reading improvement; and the teacher applies decision rules to the slope to formulate instructional decisions.

These strategies for characterizing reading competence in one timeframe and for describing growth over time using CBM have been shown to be more sensitive to inter- and intra-individual differences than those offered by other commercial tests and by other classroom reading assessments (e.g., Marston, Fuchs, & Deno, 1986). In addition, CBM is sensitive to growth made under a variety of treatments (Fuchs, Fuchs, & Hamlett, 1989b; Hintze & Shapiro, 1997; Hintze, Shapiro, & Lutz, 1994; Marston et al., 1986). In a related way, special educators' instructional plans, developed in response to CBM, incorporate a wide range of reading methods including, for example, decoding instruction, repeated readings, vocabulary instruction, story grammar exercises, and semantic mapping activities (Fuchs, Fuchs, Hamlett, & Ferguson, 1992). So, CBM is not tied to any particular reading instructional method.

Perhaps most important, however, studies indicate that CBM progress monitoring enhances special educators' capacity to plan programs for and effect achievement among students with serious reading problems. The methods by which CBM informs reading instruction rely on the graphed performance indicator. Decisions are based on a graph displaying time on the horizontal axis and performance (number of words read correctly from text in 1 min) on the vertical axis. If a student's growth trajectory is judged to be adequate, the teacher increases the student's goal for year-end performance; if not, the teacher revises the instructional program. Research shows that these decision rules produce more varied instructional programs, which are more responsive to individual needs (Fuchs et al., 1989b), with more ambitious student goals (Fuchs, Fuchs, & Hamlett, 1989a) and stronger end-of-year scores on commercial reading tests (e.g., Fuchs, Deno, & Mirkin, 1984; Wesson, 1991).

In reading, the vast majority of CBM research centers on this passage reading fluency task. When considering the developmental span within the elementary school years, however, the CBM passage reading fluency task is incomplete. The focus of the present study is the early phase of first grade, when the CBM passage reading fluency task produces a floor effect for many students. These students may make progress toward becoming a reader even though performance on the CBM passage reading fluency task, which hovers around zero, indicates a lack of growth. For this reason, alternative tasks, which are sensitive to beginning first-grade reading development, are required.

RESEARCH ON CBM MEASURES FOR BEGINNING FIRST-GRADE READING DEVELOPMENT

Prior work provides the basis for supposing that two measures may be potentially useful for indexing and monitoring beginning first-grade reading development: word identification fluency and nonsense word fluency. *Word identification fluency* is one of the CBM measures investigated by Deno and colleagues (Deno, Mirkin, & Chiang, 1982) at the University of Minnesota Institute for Research on Learning Disabilities. With word identification fluency, students have 1 min to read isolated words, presented in lists; the words are selected randomly from high frequency word lists. The score, the number of words read correctly, represents automatic word recognition skill, a hallmark of competent reading behavior. Deno et al. examined the concurrent validity of word identification fluency with 66 children in Grades 1 to 6. The criterion measures were the reading comprehension subtest of the Peabody Individual Achievement Test and the phonetic analysis and the inferential and literal reading comprehension subtests of the Stanford Diagnostic Reading Test. When students read third-grade word lists, correlations with these four respective measures were .76, .68, .71, and .75; when they read sixth-grade word lists, respective correlations were .78, .71, .68, and .74.

Nonsense word fluency is the first-grade Dynamic Indicators of Basic Literacy Skills, or DIBELS, measure developed more recently by Roland Good, Ruth Kaminski, and colleagues at the University of Oregon (Good, Simmons, & Kame'enui, 2001). With nonsense word fluency, students have 1 min to read consonant-vowel-consonant pseudowords. The score is the number of sounds produced correctly, with credit earned either by saying individual sounds in the pseudowords or by phonologically recoding the pseudowords (with three sounds awarded for each correctly read pseudoword). The score indexes "letter-sound correspondence and the ability to blend letters into words in which letters represent their most common sounds" (pp. 272-273). As Good et al. showed with samples of 70 to 242 children, concurrent validity with the Woodcock-Johnson readiness cluster score (i.e., visual auditory learning and letter identification) ranged between .35 in May to .59 in February (median coefficient = .52). The predictive validity coefficients from October of first grade to May of first grade were .71 with respect to CBM passage reading fluency and .52 with respect to the Woodcock-Johnson reading cluster score.

PURPOSE AND IMPORTANCE OF THE PRESENT STUDY

Traditional concurrent and predictive validity, as demonstrated by the Deno et al. (1982) and Good et al. (2001) research groups, are important. By documenting correspondence between the CBM task and important criterion variables, criterion validity verifies that a test measures relevant behaviors and helps establish that the measure can be used to identify students with serious reading problems. Yet, Deno et al.'s findings were restricted to concurrent validity and did not focus on first graders reading first-grade word lists, and the Good et al. concurrent validity coefficients' criterion measures did not involve reading. Moreover, although concurrent and predictive validity are necessary features of an adequate progress-monitoring task, they do not constitute sufficient evidence that a measure works well for the purpose of monitoring progress over time. Some measures may function well as correlates or predictors of important criterion outcomes (when those measures are collected at one point in time), but fail to represent reading improvement when administered frequently for progress monitoring.

At the same time, technical features of these early reading measures (word identification fluency and nonsense word fluency) have never been contrasted with the same sample using the same procedures. Therefore, direct comparisons of these two alternative CBM measures are not possible. In contrast, the CBM literature, by convention, offers direct comparisons of the tenability of alternative progress-monitoring tasks (e.g., Deno et al., 1982; Fuchs & Fuchs, 1992) so that progress-monitoring measures may be compared and optimal measures identified. This is important because no absolute criterion exists for judging adequate criterion validity. Instead, "our basis for evaluating any one predictor is in relation to other possible predictors" (Thorndike & Hagen, 1969, p. 169). So, direct comparison between word identification fluency and nonsense word fluency is required.

We designed the present study to fill these gaps in the CBM literature on early progress monitoring in the area of reading. Toward that end, we contrasted the concurrent and predictive validity for the two alternative CBM early reading measures: word identification fluency and nonsense word fluency. For predictive validity, we investigated both CBM performance level and CBM slope of improvement. Whereas the predictive validity of performance level (in one timeframe) is relevant for judging the validity of a measure for screening, the predictive validity of slope (improvement over time) is necessary for evaluating the validity of a measure for progress monitoring.

We assessed a relatively large cohort ($n = 151$) of at-risk children (the target group for progress monitoring) in the fall of first grade on the two early reading CBM measures and then again in the spring; at each time point, we administered criterion measures that directly involved reading (the Woodcock Reading Mastery Tests Word Identification and Word Attack subtests at both time points; the Comprehensive Reading Assessment Battery, which assesses passage reading fluency and comprehension, added in the spring). In addition, we assessed each of the 151 children for 20 weeks, once weekly for the first 7 weeks and twice weekly for the final 13 weeks, on the two alternative early reading CBM measures.

Findings are potentially important in helping school personnel identify useful measures for monitoring the progress of children as they enter and progress through first grade. As schools struggle with No Child Left Behind's requirement to monitor the reading progress of children in kindergarten through third grade, this topic takes on a sense of urgency. As special educators strive to infuse greater accountability to the individualized education program (IEP) progress-monitoring mandate, findings are equally useful for special educators. Across general and special education, CBM users need to have confidence that the measures they use to identify students for expensive, intensive services, in fact, yield the stu-

Across general and special education, CBM users need to have confidence that the measures they use to identify students for expensive, intensive services, in fact, yield the students most at risk for poor reading outcomes.

dents most at risk for poor reading outcomes. Practitioners also need to be assured that when students' CBM scores are increasing over time, those CBM slopes are associated with improved scores on high-stakes tests and other important outcomes that bespeak adequate end-of-first grade reading.

METHOD

PARTICIPANTS

The sample for the present set of analyses was derived from an intervention study examining the effects of Peer-Assisted Learning Strategies (PALS) in first grade (McMaster, Fuchs, Fuchs, & Compton, in press). For that study, eight schools were identified in a large, Southeastern, metropolitan public school system; in each school, at least three first-grade teachers volunteered to participate. Half the schools were high poverty (i.e., Title I), and half were designated as non-Title I. In these schools, we identified 33 teachers for participation and, blocking within school, randomly assigned teachers to control (one third of the teachers) or PALS (two thirds of the teachers). Children in the control classrooms were not included in the present database because we did not monitor their performance with either CBM measure; therefore, control students had no data to contribute to the analyses described in this article.

In PALS classrooms, we screened all children on whom we had parental consent using rapid letter naming (number of letters named in 1 min in response to 26 randomly displayed, lowercase letters). Using these scores, we designated the lowest 7 children per class (i.e., 154 children in all) as at risk for reading difficulties. We assessed these students in the spring and fall using the two CBM measures as well as a set of criterion reading measures. We also monitored these at-risk children for 20 weeks, once weekly for the first 7 weeks and twice weekly for the final 13 weeks, with the two CBM measures. The 151 children on whom we had complete data constituted the sample for the present analyses.

On average, the age of these children in October of first grade was 6.72 years ($SD = 0.46$). Of these 151 children, 74 were in Title I schools and 77 were in non-Title I schools; 89 received subsidized lunch; 79 were male; 26 were English language learners; 11 had IEPs (6 had speech/language disability, and 5 had a learning disability); 17 had been retained (7 in kindergarten and 10 in first grade); and 58 were African American, 53 were European American, 36 were Hispanic, and 4 were Asian. During the fourth 6-week marking period, 63 children had perfect attendance, 27 had 1 absence, 28 had 2 absences, 11 had 3 absences, 8 had 4 absences, 6 had 5 absences, 3 had 6 absences, 1 each had 7, 8, and 11 absences, and 2 had 13 absences.

CRITERION MEASURES

Word Attack Subtest of the Woodcock Reading Mastery Test-Revised, Form G (Woodcock, 1987). This measure evaluates students' ability to pronounce pseudowords. It contains 45 nonsense words, ordered from most easy to most difficult. The test is discontinued after six consecutive errors. Students earn 1 point for each correctly pronounced pseudoword. Scores range from 0 to 45. Split-half and test–retest reliabilities are .95 and .90, respectively, for first grade. Concurrent validity with respect to the Woodcock Johnson at first grade was .57 with Letter-Word Identification, .64 with Word Attack, .43 with Passage Comprehension, and .69 with Total Reading.

Word Identification Subtest of the Woodcock Reading Mastery Test-Revised, Form G (Woodcock, 1987). The Word Identification subtest requires children to read single words. It consists of 100 words ordered in difficulty. Testing is discontinued after six consecutive errors. Students earn 1 point for each correctly pronounced word. Scores range from 0 to 100. Split-half and test–retest reliabilities are .99 and .94, respectively, for first grade. Concurrent validity with respect to the Woodcock Johnson at first grade was .69 with Letter-Word Identification, .48 with Word Attack, .75 with Passage Comprehension, and .82 with Total Reading.

Comprehensive Reading Assessment Battery (CRAB). The CRAB (Fuchs, Fuchs, & Hamlett, 1989c) employs 400-word traditional folktales used in previous studies of reading comprehension (e.g., Brown & Smiley, 1977; Jenkins, Heliotis, Haynes, & Beck, 1986). The folktales had been rewritten by Jenkins et al. to approximate a second- to-third-grade readability level (Fry, 1968), while preserving the gist of the stories. We rewrote them at a readability grade level of 1.5. Students first read aloud one folktale for 3 min and then answer 10 comprehension questions. This is repeated on a second folktale. The questions, developed by Jenkins et al., require short answers, reflecting recall of information contained in idea units of high thematic importance. To generate the fluency score, examiners mark insertions, omissions, mispronunciations, hesitations of more than 4 s, and substitutions as errors (self-corrections are not considered errors). The score is the average number of correct words-read-across the two passages. In this study, we divided the score by 3 min to produce a words-read-correctly per-minute metric. Test–retest reliability ranges from .93 to .96; concurrent validity with the reading comprehension subtest of the Stanford Achievement Test (SAT) was .91 (Fuchs, Fuchs, & Maxwell, 1988). To generate the comprehension score, the tester records student answers to the comprehension questions. When the student makes five consecutively incorrect responses, questioning is terminated. The score is the average number of questions answered correctly across the two passages. For the number of questions answered correctly score, test–retest reliability was .92; the correlation with the SAT was .82 (Fuchs et al., 1988). On the present sample, correlations for the CRAB fluency score were .81 with Woodcock Word Identification and .58 with Woodcock Word Attack; for the CRAB comprehension score, .71 with Woodcock Word Identification and .59 with Woodcock Word Attack.

PROGRESS-MONITORING MEASURES

Word Identification Fluency. With word identification fluency, the child is presented with a single page of 50 high-frequency words. Alternate forms were generated by randomly sampling words, with replacement, from 100 high frequency words from the Dolch preprimer, primer, and first-grade-level lists. The student has 1 min to read words. If a student hesitated on an item for 4 seconds, the examiner prompted him/her to proceed to the next word. The alternate test-form/stability coefficient from 2 consecutive weeks was .97; from 2 consecutive months, .91 (Fuchs, Compton, Fuchs, & Bryant, 2004). As calculated on the current sample, alternate test-form/stability from 2 consecutive weeks was .88.

Nonsense Word Fluency. With nonsense word fluency, the child is presented with a single page of 50 consonant-vowel-consonant or vowel-consonant pseudowords. Alternate forms were printed from the DIBELS Web site (http://dibels.uoregon.edu). The student has 1 min to say the sounds constituting the pseudowords or to read the pseudowords. If a student lingered on an item for 4 s, the examiner prompted him/her to proceed to the next item. The score is the number of correctly spoken sounds (with three sounds awarded for a correctly read pseudoword). The median alternate test-form/stability coefficient at 2-month intervals was .83 (Good et al., 2001). As calculated on the current sample, alternate test-form/stability from 2 consecutive weeks was .87.

DATA COLLECTION

Data collectors were full-time master's students, full-time doctoral students, or full-time employees with master's degrees, all of whom had been trained to 100% accuracy in data-collection procedures prior to any data collection. They administered tests in quiet locations in the schools, working with one child at a time. Interscorer agreement, calculated on 20% of protocols by two independent scorers, ranged between 98% and 100%. The fall battery of criterion measures (Woodcock Word Identification and Word Attack) was administered in October. The spring battery of criterion measures (Woodcock Word Identification and Word Attack plus Comprehensive Reading Assessment Battery) was administered in May. Progress-monitoring measures were administered for 20 weeks, once weekly for the first 7 weeks and twice weekly for the final 13 weeks, with word identification fluency and nonsense word fluency administered in the same session and with make-ups completed within 1 week of the targeted data-collection date.

DATA ANALYSIS AND RESULTS

To obtain a *level of performance* on each progress-monitoring measure, two measurements were averaged for the fall (first two measurements) and for the spring (last two measurements). To obtain

TABLE 1
Means and Standard Deviations (n = 151)

	Occasions					
	Fall		Spring		Year	
Variable	X	(SD)	X	(SD)	X	(SD)
Criterion Variables						
WRMT-R WID	9.01	(7.98)	30.92	(11.43)	NA	
WRMT-R WAT	3.27	(4.37)	11.64	(6.87)	NA	
CRAB Fluency	NA		29.65	(22.17)	NA	
CRAB Comp	NA		1.33	(1.58)	NA	
CBM Level						
NWF	31.29	(14.47)	52.17	(24.07)	NA	
WIF	10.11	(9.26)	29.72	(19.96)	NA	
CBM Slope						
NWF	1.92	(2.04)	1.49	(1.23)	1.24	(0.82)
WIF	0.90	(0.90)	1.31	(0.95)	1.02	(0.68)

Note: WRMT-R is Woodcock Reading Mastery Test-Revised; WID is the Word Identification subtest; WAT is the Word Attack subtest; CRAB is the Comprehensive Reading Assessment Battery; Fluency is number of words read correctly aloud per minute; Comp is comprehension questions answered correctly; WIF is word-identification fluency; NWF is nonsense word fluency.

a *slope of improvement* on each progress-monitoring measure, an ordinary least-squares regression was calculated between calendar days and scores. That slope was converted to a weekly slope (by multiplying the derived slope by 7 days) to represent the weekly increase in score. Slopes were derived for fall (October through December), spring (January through May), and the year (October through May).

Means and standard deviations for these scores and for the criterion variables are shown in Table 1. Mean raw scores on the Woodcock measures correspond to near-average standard scores. This is surprising given that this at-risk sample represented the lowest third of the classrooms in an urban setting. An inflated normative profile at first grade on the Woodcock Reading Mastery Tests has, however, been documented elsewhere (e.g., Fuchs, et al., 2004; Vellutino, Scanlon, Small, & Fanuele, 2003), and evidence that this sample was, in fact, at risk is provided in other ways. That is, relative to a normative profile of 179 children representing the full range of first-grade performance (e.g., Fuchs et al., 2004), the fall CBM word identification fluency score in the present study is .85 standard deviations below the mean and the spring CBM word identification fluency score in the present study is 1.24 standard deviations below the mean. Moreover, the end-of-year CRAB fluency score is considerably below a benchmark performance of 40, even though these children participated in the research-validated PALS reading treatment.

We next ran a series of correlations between word identification fluency and the criterion variables and between nonsense word fluency and the criterion variables. Each pair of correlations (word identification fluency vs. nonsense word fluency) was compared using Walker and Lev's (1953) formula, which tested the difference between correlations calculated on dependent samples.

To index *concurrent validity*, we ran correlations between the fall progress-monitoring level and the fall criterion measures (Woodcock Word Identification and Word Attack) and between the spring progress-monitoring level and the spring criterion measures (Woodcock Word Identification and Word Attack; CRAB Fluency and Comprehension). These correlations are reported in the top six rows of Table 2. As these *t* values in Table

TABLE 2

Concurrent and Predictive Validity for Word Identification Fluency Versus Nonsense Word Fluency in First Grade (n = 151)

Validity	CBM Measure		
	WIF	NWF	t(148)ª
Concurrent Validity			
Fall CBM Level			
WRMT-R WID	.77	.58	4.93***
WRMT-R WAT	.59	.50	1.12
Spring CBM Level			
WRMT-R WID	.82	.64	3.82***
WRMT-R WAT	.52	.51	0.21
CRAB Fluency	.93	.80	2.72**
CRAB Comprehension	.73	.54	3.23**
Predictive Validity			
Fall CBM Level			
WRMT-R WID	.63	.57	1.26
WRMT-R WAT	.45	.46	0.19
CRAB Fluency	.80	.64	4.27***
CRAB Comprehension	.66	.50	5.79***
Fall CBM Slope			
WRMT-R WID	.43	.05	3.96***
WRMT-R WAT	.27	-.03	2.93**
CRAB Fluency	.54	.16	4.27***
CRAB Comprehension	.49	-.04	5.71***
Spring CBM Slope			
WRMT-R WID	.61	.35	3.52***
WRMT-R WAT	.32	.27	0.79
CRAB Fluency	.63	.49	2.08*
CRAB Comprehension	.45	.27	2.13*
Year CBM Slope			
WRMT-R WID	.79	.38	8.18***
WRMT-R WAT	.50	.28	3.83***
CRAB Fluency	.85	.58	6.84***
CRAB Comprehension	.66	.27	6.80***

Note: WIF is word-identification fluency; NWF is nonsense word fluency; WRMT-R is Woodcock Reading Mastery Test-Revised; WID is the Word Identification subtest; WAT is the Word Attack subtest; CRAB is the Comprehensive Reading Assessment Battery.

ªTo test the difference between correlations calculated on the same sample, we relied on Walker and Lev's (1953) formula.

*$p < .05$. **$p < .01$. ***$p < .001$.

TABLE 3
Predictive Validity for Fall Woodcock Versus Fall CBM Measures in First Grade (n = 151)

Spring Criterion	Fall Predictor				t(148)[a]			
	WRMT-R							
	WID	WAT	WIF	NWF	WID v. WIF	WID v. NWF	WAT v. WIF	WAT v. NWF
WRMT-R WID	.60	.44	.63	.57	0.71	0.53	**3.30**	1.98
WRMT-R WAT	.46	.49	.45	.46	0.20	0.00	0.63	0.43
CRAB-F	.63	.47	.80	.64	**5.01**	0.19	**7.46**	**2.76**
CRAB-C	.59	.55	.66	.66	1.70	0.18	**2.05**	0.76

Note: WRMT-R is Woodcock Reading Mastery Test-Revised; WID is the Word Identification subtest; WAT is the Word Attack subtest; WIF is word-identification fluency; NWF is nonsense word fluency; CRAB is the Comprehensive Reading Assessment Battery; F is the CRAB passage reading fluency score; C is the CRAB comprehension score.

[a] Using Walker and Lev's (1953) formula, these *t* tests compare correlations between the top fall predictor versus the bottom fall predictor in the relevant column with respect to spring criterion in the relevant row (e.g., the first *t* test compares the correlation between fall WRMT-R WID and spring WRMT-R WID to the correlation between fall word identification fluency and spring WRMT-R WID). For 148 degrees of freedom, the *t* value associated with $p < .05$ is 1.976; the *t* value associated with $p < .01$ is 2.61; and the *t* value associated with $p < .001$ is 3.36. Any *t* value associated with $p < .05$ is bolded.

2 reveal, correlations for the CBM word identification fluency measure were reliably higher than for nonsense word fluency for one of the two fall concurrent validity criterion variables and for three of the four spring concurrent validity criterion variables. Basically, comparisons favored word identification fluency except when the criterion variable was highly aligned with nonsense word fluency (i.e., Woodcock Word Attack). Even there, where we would expect the comparison to favor nonsense word fluency, concurrent validity for the two CBM measures was, in fact, comparable.

To index *predictive validity*, the following indices were correlated with the spring criterion measures (Woodcock Word Identification and Word Attack; CRAB Fluency and Comprehension): fall progress-monitoring level (see rows 7–10 of Table 2), fall progress-monitoring slope (see rows 11–14 of Table 2), spring progress-monitoring slope (see rows 15–18 of Table 2), and the year's progress-monitoring slope (see rows 19–22 of Table 2). As with concurrent validity, the vast majority of predictive validity coefficients reliably favored word identification fluency over nonsense word fluency.

We then looked at how the predictive validity of the two CBM measures compared to the predictive validity of the two Woodcock measures. These correlations are shown in Table 3. When spring Woodcock Word Attack was the criterion variable, all four predictors performed comparably. Nonsense word fluency performed comparably to the Woodcock measure when predicting Woodcock Word Identification and when predicting CRAB comprehension, but outperformed both Woodcock measures when predicting CRAB fluency. By contrast, word identification fluency reliably outperformed both Woodcock measures in predicting Woodcock Word Identification, in predicting CRAB fluency, and in predicting CRAB comprehension.

Finally, we performed dominance analysis (Budescu, 1993; Schatschneider, Francis, Fletcher, & Foorman, 2004), which is an extension of multiple regression. Dominance analysis involves the pairwise comparison of all predictors (i.e., fall nonsense word fluency level, full-year nonsense word fluency slope, fall word identification fluency level, full-year word identification fluency slope) as they relate to a spring criterion (i.e., Woodcock Word Identification, Woodcock Word Attack, CRAB fluency, CRAB comprehension). With dominance analysis, a variable is considered dominant over another if the predictive ability of

TABLE 4
Dominance Analysis of the Predictors of Word Recognition, Word Attack, Passage Reading Fluency, and Comprehension (n = 151)

	\multicolumn{12}{c}{Outcome}											
	WRMT-WID			WRMT-WAT			CRAB-F			CRAB-C		
Predictors	R^2D	Asy. SE	p	R^2D	Asy. SE	p	R^2D	Asy. SE	p	R^2D	Asy. SE	p
NWF-L v. WIF-L	-.03	.02	ns	.01	.03	ns	**-.12**	**.02**	**< .05**	**-.13**	**.04**	**< .05**
NWF-L v. NWF-S	-.00	.04	ns	.02	.05	ns	-.01	.02	ns	-.03	.04	ns
NWF-L v. WIF-S	**-.21**	**.05**	**< .05**	.03	.05	ns	**-.13**	**.03**	**< .05**	**-.14**	**.06**	**<.05**
WIF-L v. NWF-S	.03	.02	ns	.01	.02	ns	**.11**	**.02**	**< .05**	**.10**	**.04**	**< .05**
WIF-L v. WIF-S	**-.18**	**.05**	**< .05**	-.03	.04	ns	-.01	.04	ns	-.01	.06	ns
NWF-S v. WIF-S	**-.21**	**.04**	**< .05**	-.05	.03	ns	**-.12**	**.03**	**< .05**	**-.11**	**.04**	**< .05**

Note: WRMT-R is Woodcock Reading Mastery Test-Revised; WID is the Word Identification subtest; WAT is the Word Attack subtest; CRAB-F is the passage reading fluency score of the Comprehensive Reading Assessment Battery; CRAB-C is the comprehension score of the Comprehensive Reading Assessment Battery. R^2D is the difference between the squared multiple correlations; Asy. SE is the standard error of the differences; p indicates whether the lower and upper bounds of the 95% asymptotic confidence interval included zero. NWF is nonsense word fluency; WIF is word-identification fluency; L is fall level; S is full-year slope. Significant differences are bolded.

that variable exceeds the other—both alone and in the presence of all other predictors in the model. Dominance analysis uses asymptotic confidence limits to test differences in the unique effects among pairwise comparisons. These pairwise comparisons are not a test of the amount of unique variance each predictor contributes, but instead are a direct comparison of the *differing amounts of unique variance attributed to the two predictors* as they relate to the spring criterion.

In Table 4, we show the asymptotic confidence intervals for each of the six pairs of predictors. Each row shows the pair of variables compared. Each column shows results of the dominance analysis for one spring outcome variable. Under each spring outcome variable, three statistics for the dominance analysis are displayed: R^2D is the difference between the squared multiple correlations; asymptotic SE is the standard error of the differences; p indicates whether the lower and upper bounds of the 95% asymptotic confidence interval include zero (see Budescu, 1993; Hedges & Olkin, 1981). If a confidence interval does not include zero, the difference in unique variances is significant at an alpha level of .05 (Budescu). For example, the first cell in Table 4 (i.e., first row and first column) compares the unique variance that nonsense word fluency level accounts for above and beyond word identification fluency level (-3%) in the presence of all four predictor variables. The negative sign indicates that word identification fluency accounts for more unique variance than nonsense word fluency (if the reverse were true, then the sign would be positive), but the p value indicates that this difference is not significant. Therefore, in predicting Woodcock Word Identification, word identification fluency level does not dominate nonsense word fluency level. In Table 4, we have highlighted the cells where the difference in unique variances are significantly different from zero and, in those cases, highlighted the predictor variable that accounts for more unique variance.

With these analyses, we were interested in whether word identification fluency dominated nonsense word fluency. We were also interested in examining whether slope provided additional predictive value over performance level indices. Word identification fluency dominated nonsense word fluency in 10 of 16 comparisons (see rows 1, 3, 4, and 6 of Table 4); word identification fluency slope provided additional predictive value over performance level indices in 4 of 8 comparisons (see rows 3 and 5 of Table 4); nonsense word fluency slope provided additional predictive value over performance level indices in 2 of 8 comparisons (see rows 2 and 4 of Table 4); and word identification fluency slope dominated nonsense word fluency slope in 3 of 4 comparisons (see last row of Table 4).

DISCUSSION

To identify optimal progress-monitoring tasks, direct comparison of various measures is necessary because no absolute criterion exists for judging adequate criterion validity. Instead, the basis for evaluating a predictor is in relation to other possible predictors (Thorndike & Hagen, 1969). Toward that end, this study compared two potentially useful CBM measures for monitoring early reading development in first grade: word identification fluency and nonsense word fluency. For these measures, we examined (a) concurrent validity for CBM level at fall and spring; (b) predictive validity for CBM level from fall to spring; and (c) predictive validity for fall CBM slope to spring final status, for spring CBM slope to spring final status, and for full-year CBM slope to spring final status. Almost all comparisons favored word identification fluency.

To explore concurrent validity near the beginning of first grade, we ran correlations between the two CBM measures and Woodcock Word Identification (which requires students to read words in untimed fashion) and Woodcock Word Attack (which requires students to decode pseudowords in untimed fashion). Given the nature of the two CBM measures, one might expect word identification fluency to correlate more strongly with Woodcock Word Identification and expect nonsense word fluency to correlate more strongly with Woodcock Word Attack. This was only partly true. The correlation with Woodcock Word Identification was, in fact, statistically significantly higher for word identification fluency than for nonsense word fluency (.77 vs. .58). Contrary to expectations, however, the correlations for the two CBM measures with Woodcock Word Attack were comparable (.59 for word identification fluency vs. .50 for nonsense word fluency). So, even at the beginning of the year, when one might assume a lower floor (and therefore greater range) for nonsense word fluency (which awards credit for saying sounds without requiring decoding), greater validity was demonstrated for the CBM word identification fluency measure.

At spring, where we included a better variety of criterion measures, tapping text reading fluency and comprehension with the CRAB, findings again supported word identification fluency over nonsense word fluency, this time even more strongly. Across Woodcock Word Identification, CRAB fluency, and CRAB comprehension, correlation coefficients ranged between .73 to .93 for word identification fluency; between .51 and .80 for nonsense word fluency. The one criterion measure for which correlations between the two CBM measures were comparable (.52 vs. .51) was the Woodcock Word Attack, where the task (reading pseudowords) is more similar to nonsense word fluency than to word identification fluency. Thus, at the end of first grade, the word identification fluency task remains a stronger concurrent correlate of important reading behaviors.

In a similar way, for predictive validity, results favored CBM word identification fluency over nonsense word fluency. Fall CBM word identification fluency scores demonstrated superior predictive validity with respect to the CRAB fluency and comprehension criterion measures. Consequently, for identifying children in October of first grade who are at risk for poor end-of-year reading outcomes, the CBM word identification fluency measure outperforms the nonsense word fluency task.

To supplement these analyses, we also asked whether we could predict spring outcomes just as well—simply by using fall Woodcock scores. The answer was no. In predicting spring Woodcock Word Identification, the fall CBM measures did comparably well to fall Woodcock Word Identification. Moreover, in predicting spring Woodcock Word Attack, both CBM measures outperformed the fall Woodcock Word Attack measure, and the significant differences in predictions of spring CRAB performance consistently favored the fall CBM measures over the fall Woodcock measures. So, in all cases, there was no advantage to using fall Woodcock scores to predict spring reading performance.

Of course, the major purpose for CBM is progress monitoring, where the criterion validity of the CBM slope is more important than CBM level. With CBM progress monitoring, slope is a critical index because CBM slope is used to formulate decisions about whether reading progress is adequate. If the CBM slope indicates adequate progress, the instructional program remains intact; if not, the program is revised. So, the relevant technical question is: Does improvement on the CBM measure, as indexed with slope, reflect meaningful reading development, which predicts reading accomplishment at the end of the year?

Across the fall semester, CBM slopes correlated statistically significantly higher for word identification fluency than for nonsense word fluency with all four spring criterion measures. In fact, coefficients for the nonsense word fluency measure slopes were disappointingly low, ranging from -.04 to .16. Because nonsense word fluency is recommended for progress monitoring in the fall of first grade within the DIBELS system (Good et al., 2001), these findings raise serious concern. An increasing pattern of scores through the first semester of first grade on DIBELS nonsense word fluency appears to bear little relationship to students' end-of-year reading status. By contrast, for word identification fluency, fall slopes of improvement correlated more strongly, with coefficients of .43 with end-of-year Woodcock Word Identification, .54 with CRAB fluency, and .49 with CRAB comprehension. These correlations are modest, but fall in the range of many predictive measures (Jensen, 1981). And the correlations are noteworthy given the difficulty of achieving strong correlations when measures of change, such as slope, are used as predictors. At this time, therefore, fall word identification fluency slope appears to represent an acceptable index for predicting end-of-year reading outcome. Clearly, it represents a better alternative than fall nonsense word fluency slope.

As might be expected, given the more proximate timeframe, spring slopes of improvement correlated with final reading status measures more strongly than did fall slopes. However, here again, coefficients for spring word identification fluency slopes were reliably stronger than for nonsense word fluency slopes. Perhaps most important, however, very large (and reliable) differences in the magnitude of correlations were observed for the full-year slopes: Coefficients for full-year nonsense word fluency slopes ranged between .27 (with CRAB comprehension) to .58 (with CRAB fluency). By contrast, coefficients for full-year word identification fluency slopes ranged from .50 (with Woodcock Word Attack) to .85 (with CRAB fluency). In addition, the superiority of word identification fluency slope (over nonsense word fluency slope) was demonstrated even for Woodcock Word Attack, which seems more transparently related to nonsense word fluency than to word identification fluency.

Dominance analysis provided a supplementary and elegant method for exploring the predictive value of the two early reading progress-monitoring measures. In each case, word identification fluency, level, or slope, dominated nonsense word fluency. Word identification fluency level dominated nonsense word fluency level in predicting CRAB fluency and CRAB comprehension; word identification fluency slope dominated nonsense word fluency level in predicting all spring outcome variables except Woodcock Word Attack; word identification fluency level dominated nonsense word fluency slope in predicting CRAB comprehension; word identification fluency slope dominated word identification fluency level in predicting Woodcock Word Identification; and word identification fluency slope dominated nonsense word fluency slope in predicting all spring outcome variables except Woodcock Word Attack. These summative analyses not only corroborate the superiority of word identification fluency, but also suggest that collecting word identification fluency slope provides additional predictive value beyond simply collecting fall word identification fluency level data. This indicates the benefit of progress monitoring beyond initial screening for identifying students likely to experience reading difficulty.

In sum, results suggest that word identification fluency functions better than nonsense word fluency as a CBM tool for assessing early reading development in first grade. Because predictive validity with respect to end-of-year text-reading fluency and comprehension is stronger for word identification fluency than for nonsense word fluency, word identification fluency provides a stronger basis for formulating screening decisions in October of first grade. Moreover, the superiority of the word identification fluency over nonsense word fluency is most clearly demonstrated for progress monitoring decisions, where 11 of 12 correlations for CBM slope with respect to end-of-year outcomes were stronger for word identification fluency than for nonsense word fluency. Dominance analysis demonstrated how slope provides additional predictive value beyond one-time screening decisions in the fall. Dominance analysis also corroborates that word identification fluency, fall level, and full-year slope dominates nonsense word fluency, when both metrics for both CBM measures are entered simultaneously to predict spring outcomes. For these reasons, practitioners can have confidence that increases in word identification fluency over time reflect improved performance on important end-of-year reading outcomes. As our results suggest, the same is not true for DIBELS nonsense word fluency, and findings are particularly compelling because data were collected on the same group of children using the same methods.

Why is predictive validity for word identification fluency performance level and slope better than for nonsense word fluency? Although findings do not yield a direct answer to this question, we offer two possible explanations, which correspond to two difficulties with the nonsense word fluency task. First, on the nonsense word fluency task, two students with very different performance patterns may receive equal credit. That is, a student who says three separate sounds in response to a consonant-vowel-consonant pseudoword earns three points—the same score as a student who blends those three sounds into the pseudoword. Clearly, the student who blends the sounds has stronger reading capacity than the child who can only represent the separate sounds. Moreover, in our sample, we observed low-performing students who, when monitored with nonsense word fluency, were increasingly capable of saying many sounds very quickly, without achieving the alphabetic insight required for blending.

A second problem is that competent phonological decoding is, especially as the year progresses, better represented by the capacity to decode a variety of phonetic patterns. So, students who perform well on nonsense word fluency's consonant-vowel-consonant pseudowords may or may not be skilled at reading consonant-vowel-consonant -*e* words, *r*-controlled words, dual vowel words, multisyllabic words, etc. The restriction of the nonsense word fluency task to a single, easy phonetic pattern may reduce the correlation between nonsense word fluency and important criterion measures.

...results suggest that word identification fluency functions better than nonsense word fluency as a CBM tool for assessing early reading development in first grade.

Present findings are reminiscent of earlier work demonstrating the superiority of word identification fluency over nonsense word fluency with a different sample of first graders with more severe reading difficulties (Fuchs, 2003). For that sample of 36 at-risk students who received one-to-one tutoring across the second semester of first grade, nonsense word fluency slopes failed to reliably discriminate student performance on key indicators of reading competence at the end of first grade. In that study, a median split was performed on the slopes of these 36 children, creating a group of children with the top 18 slopes and another group with the bottom 18 slopes. The average effect size comparing these two groups of children on end-of-year indicators of reading competence and on fall-to-spring reading growth was .4 standard deviations, and the difference in the performance of these two groups was statistically significantly different on only one of the eight criterion measures. By contrast, when top versus bottom groups were formed on the basis of word identification fluency slopes, the average effect size comparing the groups exceeded 1 standard deviation, and the performance of students with the top-half slopes versus those with bottom-half slopes were statistically significantly different on all eight year-end indicators of reading competence and fall-to-spring reading growth. Current findings corroborate those earlier findings, showing how improvement across time on word identification fluency functions better than nonsense word fluency for forecasting end-of-first-grade reading status (as well as reading improvement).

It is important to note that this study employed a restricted sample of at-risk pupils and that we might expect correlations generally to be higher if we were to conduct the study with a greater range of performance. Of course, this renders the large correlations for word identification fluency even more impressive, but as with any study, results should be corroborated with additional samples.

Nevertheless, findings are particularly timely given the press to implement the progress-monitoring component of No Child Left Behind and as special educators ratchet up the IEP progress-monitoring requirement of the Individuals with Disabilities Education Act. Moreover, given widespread adoption of DIBELS, practitioners may wish to reconsider nonsense word fluency in favor of the CBM word identification fluency measure. Clearly, as schools select measures for monitoring the reading progress of children with and without disabilities in the early stages of reading development, results provide a strong basis for selecting word identification fluency over nonsense word fluency. Findings also indicate that monitoring student progress frequently with word identification fluency can contribute importantly, beyond the simple collection of fall screening data, to the identification of students likely to experience difficulty in learning to read in the first grade.

REFERENCES

Biemiller, A. (1977–1978). Relationship between oral reading rates for letters, words, and simple text in the development of reading achievement. *Reading Research Quarterly, 13*, 223-253.

Brown, A. L., & Smiley, S. S. (1977). Rating the importance of structural units of prose passages: A problem of meta-cognitive development. *Child Development, 48*, 1-8.

Budescu, D. V. (1993). Dominance analysis: A new approach to the problem of relative importance of predictors in multiple regression. *Psychological Bulletin, 114*, 542-551.

Deno, S. L. (1985). Curriculum-based measurement: The emerging alternative. *Exceptional Children, 52*, 219-232.

Deno, S. L., Mirkin, P. K., & Chiang, B. (1982). Identifying valid measures of reading. *Exceptional Children, 49*, 36-45.

Fry, E. B. (1968). A readability formula that saves time. *Journal of Reading Behavior, 11*, 513-516.

Fuchs, D., Compton, D. L., Fuchs, L. S., & Bryant, J. D. (2004). *Identifying students with learning disabilities using a response-to-instruction approach in first grade.* Manuscript in preparation.

Fuchs, D., & Fuchs, L. S. (1992). Limitations of a feel-good approach to consultation. *Journal of Educational and Psychological Consultation, 3*(2), 93-97.

Fuchs, L. S. (2003). Assessing intervention responsiveness: Conceptual and technical issues. *Learning Disabilities Research & Practice, 18*, 172-186.

Fuchs, L. S., & Deno, S. L. (1991). Paradigmatic distinctions between instructionally relevant measurement models. *Exceptional Children, 57*, 488-501.

Fuchs, L. S., Deno, S. L., & Mirkin, P. K. (1984). The effects of frequent curriculum-based measurement and evaluation on pedagogy, student achievement, and student awareness of learning. *American Educational Research Journal, 21*, 449-460.

Fuchs, L. S., Fuchs, D., & Hamlett, C. L. (1989a). Effects of alternative goal structures within curriculum-based measurement. *Exceptional Children, 55*, 429-438.

Fuchs, L. S., Fuchs, D., & Hamlett, C. L. (1989b). Effects of instrumental use of curriculum-based measurement to enhance instructional programs. *Remedial and Special Education, 10*(2), 43-52.

Fuchs, L. S., Fuchs, D., & Hamlett, C. L. (1989c). Monitoring reading growth using student recalls: Effects of two teacher feedback systems. *Journal of Educational Research, 83*, 103-111.

Fuchs, L. S., Fuchs, D., Hamlett, C. L., & Ferguson, C. (1992). Effects of expert system consultation within curriculum-based measurement, using a reading maze task. *Exceptional Children, 58*, 436-450.

Fuchs, L. S., Fuchs, D., Hosp, M. K., & Jenkins, J. R. (2001). Oral reading fluency as an indicator of reading competence: A theoretical, empirical, and historical analysis. *Scientific Studies of Reading, 5*, 239-256.

Fuchs, L. S., Fuchs, D., & Maxwell, L. (1988). The validity of informal reading comprehension measures. *Remedial and Special Education, 9*(2), 20-29.

Good, R. H., III, Simmons, D. C., & Kame'enui, E. J. (2001). The importance and decision-making utility of a continuum of fluency-based indicators of foundational reading skills for third-grade high-stakes outcomes. *Scientific Studies of Reading, 5*, 257-288.

Hedges, L. V., & Olkin, I. (1981). The asymptotic distribution of commonality components. *Psychometrika, 46*, 331-336.

Hintze, J. M., & Shapiro, E. S. (1997). Curriculum-based measurement and literature-based reading: Is curriculum-based measurement meeting the needs of changing reading curricula? *Journal of School Psychology, 35*, 351-375.

Hintze, J. M., Shapiro, E. S., & Lutz, G. (1994). The effects of curriculum on the sensitivity of curriculum-based measurement in reading. *The Journal of Special Education, 28*, 188-202.

Jenkins, J. R., Heliotis, J., Haynes, M., & Beck, K. (1986). Does passive learning account for disabled readers' comprehension deficits in ordinary reading situations? *Learning Disability Quarterly, 9*, 69-75.

Jensen, A. R. (1981). *Straight talk about mental tests*. New York: Free Press.

Marston, D., Fuchs, L. S., & Deno, S. L. (1986). Measuring pupil progress: A comparison of standardized achievement tests and curriculum-related measures. *Diagnostique, 11*, 77-90.

McDonnell, L. M., McLaughlin, M. J., & Morison, P. (1997). *Educating one and all: Students with disabilities and standards-based reform*. Washington, DC: National Academic Press.

McMaster, K. N., Fuchs, D., Fuchs, L. S. & Compton, D. L. (in press) Responding to nonresponders: An experimental field trial of identification and intervention methods. *Exceptional Children*.

President's Commission on Excellence in Special Education. (2002). *A new era: Revitalizing special education for children and their families*. Washington, DC: Author.

Schatschneider, C., Francis, D. J., Fletcher, J. M., & Foorman, B. R. (2004). Kindergarten prediction of reading skills: A longitudinal comparison. *Journal of Educational Psychology, 96*, 265-282.

Thorndike, R. L., & Hagen, E. (1969). *Measurement and evaluation in psychology and education* (3rd ed.). New York: John Wiley.

Vellutino, F. R., Scanlon, D. M., Small, S., & Fanuele, D. (2003, December). *Response to intervention as a vehicle for distinguishing between reading disabled and non-reading disabled children: Evidence for the role of kindergarten and first grade intervention*. Paper presented at the National Research Center on Learning Disabilities Response-To-Intervention Symposium, Kansas City, MO.

Walker, H. M., & Lev, J. (1953). *Statistical inference*. New York: Holt & Co.

Wesson, C. L. (1991). Curriculum-based measurement and two models of follow-up consultation. *Exceptional Children, 57*, 246-257.

Woodcock, R. W. (1987). *Woodcock Reading Mastery Tests* (Rev. ed.). Circle Pines, MN: American Guidance Service.

ABOUT THE AUTHORS

LYNN S. FUCHS (CEC #185), Nicholas Hobbs Professor; **DOUGLAS FUCHS** (CEC #185), Nicholas Hobbs Professor; and **DONALD L. COMPTON** (Tennessee Federation), Assistant Professor, Special Education, Peabody College of Vanderbilt University, Nashville, Tennessee.

Inquiries should be addressed to Lynn S. Fuchs, 328 Peabody, Vanderbilt University, Nashville, TN 37203.

The research described in this paper was supported in part by Grant #H324C000022 from the U.S. Department of Education, Office of Special Education Programs, and Grant HD 15052 from the National Institute of Child Health and Human Development to Vanderbilt University. Statements do not reflect the position or policy of these agencies, and no official endorsement by them should be inferred.

Manuscript received October 2003; accepted January 2004.

Online issues of the *Journal of the American Medical Association (JAMA)*

The Women's Low-Fat Diet Study vs. The Media

The February 8, 2006, issue of the *Journal of the American Medical Association* reported on some results from the Women's Health Initiative study on the benefits of a low-fat diet for women. You can access the articles online for free by going to
http://jama.ama-assn.org/contents-by-date.0.dtl
Then select 2006 and then select Feb 8. The first three articles in this issue deal with the effect of a low fat diet on different diseases. You can view the article by clicking either Full Text or PDF after the end of the citation. You will be asked to sign in, but you can choose to register for free content.

These articles generated considerable coverage in the popular press. The reports were cited on TV, radio, and in newspapers. Some of the newspaper articles are listed below. You might want to use Google instead of typing in the entire link below. Note that some of the links extend several lines. Type the entire link into your browser.

The Washington Post
http://www.washingtonpost.com/wp-dyn/content/article/2006/02/07/AR2006020701681.html

The New York Times
http://www.nytimes.com/2006/02/07/health/07cnd-fat.html?ei=5089&en=de984ad140dc68d9&ex=1296968400&partner=rssyahoo&emc=rss&pagewanted=print

Fox News
http://www.foxnews.com/story/0,2933,184409,00.html

CBS News
http://www.cbsnews.com/stories/2006/02/08/earlyshow/health/main1294374.shtml

If these links no longer work, you might use Google to find other links to this story. Be sure the date is around February 8 because there was another study on low fat diets (with somewhat different goals and results).

The objective here is to compare and contrast what the articles in the popular press say with what was actually in the study.

> The *Washington Post* article begins by stating flatly:
> *Low-fat diets do not protect women against heart attacks, strokes, breast cancer or colon cancer, a major study has found, contradicting what had once been promoted as one of the cornerstones of a healthy lifestyle.*
>
> The *New York Times* article begins with:

The largest study ever to ask whether a low-fat diet reduces the risk of getting cancer or heart disease has found that the diet has no effect.

Fox News said:
The widely believed notion that low-fat diets are good for your health went "poof" this week—although the busting of that myth shouldn't be news to regular readers of this column.

Low-fat diets didn't reduce the risk of cardiovascular disease, colorectal cancer or invasive breast cancer, according to three large studies published this week in the Journal of the American Medical Association.

The CBS News story has the headline "Study: Low-Fat Diet Big Letdown"

1. Is that what the studies showed? What would be involved in a study that showed that this diet did not protect against the diseases listed? What is the difference between saying that the study showed that there was no protection and saying that the study did not show there was protection?
2. The first of the three articles in *JAMA* deals with the risk of breast cancer. For simplicity, we will deal with that article. Similar issues can be found in the other two articles.
3. Table 1 on p. 631 is the comparison of those who used the low-fat diet (Intervention) and those who did not (Comparison). It contains a column of *P*-values. What test was used to compute the *P*-values? What is the importance of the *P*-values? Are *P*-values the best way to convey what they are trying to convey? What is the impact of the quite large sample sizes on the interpretation of the *P*-values?
4. Table 5 on p. 635 deals with the incidence of cancer. Breast cancer incidence refers to the number of subjects who developed breast cancer during the study. The line labeled Mortality deals with those who died from breast cancer during the study. The Unweighted *P*-value for Incidence is shown as 0.07. How was this computed? You should be able to perform the calculation using methods from your course. Did the authors use a one-sided or a two-sided test? How would their results have changed if they had used the other number of sides? What number of sides do you think they should have used? If you were now writing the newspaper article concerning the study results, how would the article be different from what was reported?
5. Comment on the other Unweighted *P*-values. What if the number of sides was different? Why do you think that the *P*-value for Total mortality is the largest?
6. The Weighted *P*-values were computed using a different method. However, we can consider the same issues for them as for the Unweighted values.
7. Table 5 also includes the Hazard Ratio (HR). The HR is the ratio of two rates – the probability per year for cancer in the diet group and the probability per year for the comparison group. We can think of this as the relative likelihood of getting cancer in each group. What would the null hypothesis be in terms of HR? The article gives confidence intervals. What is the main thing to look for in these

175

intervals? How is the interval for cancer incidence related to the *P*-value for cancer incidence? Why is the interval for cancer mortality so much wider than for cancer incidence? Relate the interval for Total mortality to the *P*-value for Total mortality.

8. Table 6 on p. 636 considers the impact of dietary characteristics with cancer risk. In the first row of the table, under percentage of energy for the Intervention group the table gives 18.8 (6.2). What are these two numbers and how were they calculated? The third column of this row gives 9.7 (6.2). How was the 9.7 calculated? (You can check by looking at other rows of the table.) Where did the 6.2 value come from? What does it represent?

9. The last column gives the Hazard Ratio between the Intervention group and the Comparison group for each of the four levels of percent energy. The last column gives a single *P*-value for all four of the hazard ratios. The footnote for this column is not very informative. What do you think they are testing in this column? Why is this important? Is there information in this *P*-value that could not be deduced from the four confidence intervals?

10. Table 8 on p. 639 also computes a single *P*-value for several hazard ratios (HR). Consider the questions above as they apply to Table 8.

"Association of Long-Distance Corridor Walk Performance with Mortality, Cardiovascular Disease, Mobility Limitation and Disability," Anne B Newman, et al, *JAMA*, 295(17):2018–2026.

Use the methods given for the previous article to obtain this article online.

The purpose of this study is to use the ability (and time required) for a 400 m corridor walk as an indicator of health in people aged 70–79.

1. Table 1 compares various characteristics among those who were excluded, those who stopped the walk and those who completed the walk in different times. The last four columns of Table 1 correspond to different quartiles of walk time.
   a. Which column corresponds to the slowest walkers?
   b. The first three columns have the same number of subjects, but the fourth column has eight more. If these are quartiles, why are there not the same number in each column?
   c. (Related to (b).) The times for the first three columns have small minus signs in them. What might these mean? If someone took exactly 323 seconds for the walk, in which column would they fall?
2. On p. 2021 under Results, the authors say that those who completed the walk were slightly younger. On what do they base this? What might a confidence interval for this look like?
3. In the same section, the authors say that those who walked faster were slightly younger. There are several ways to approach this. Discuss the pros and cons of a few of the approaches.
4. In Table 2, the authors compare the rates of various negative outcomes. Consider the Mortality for those who Completed the study. In the third column, it says that the Events per 1000 Person-Years is 24.7. Compare this to what you get if you divide 284 by the total Per-Years (found by multiply 2324 by 6). Why doesn't this calculation yield 24.7? What additional information would you need to compute the 24.7 figure?
5. The last three columns in Table 2 deal with Hazard Ratios (HR). This is the ratio of the hazard rates for two groups. In Table 2, the denominator is computed based on those who Completed the walk. We can think of the hazard rate as the probability of this negative outcome in a fixed time period, say, a year. The values in these three columns generally range from a bit over 1.0 up to 2.5, or even slightly over 3. What is the null hypothesis in this problem? What values of the HR would lead you to reject H0?
6. The last three columns of Table 2 are all adjusted in various ways. How this adjustment was done is almost surely beyond the scope of your course. It is sufficient to say that factors other than the ability to complete the walk would have an effect on the particular negative outcome. Consider the 1.74 figure in the first row. If those who were Excluded were older or have a different gender makeup than those who Completed, this would account for the difference in death rates. A calculation has been done to determine the death risk based on age and sex. The 1.74 figure is the additional risk after considering age and sex

differences between the two groups. If we did not do the adjustment, then we could simply compare the values in the third column of Table 2. How much did the adjustment alter the comparisons? Is doing the adjustment a good idea or not?

7. The confidence intervals for the Hazard Ratio can also tell us if there is any difference in the HR between various groups. In what cases could we say that there is a significant difference in the risks? There are 24 intervals in Table 2. Does the fact that there are quite a few intervals given complicate determining which differences are significant.

8. In the third column on p. 2021, the authors discuss Table 3, which gives the Hazard Ratio per minute of time needed for the walk. Near the end of the paragraph, they say, "the HR for persistent mobility limitation was higher than observed for mortality or CVD." What does this mean? Is this surprising? Is this a major result?

9. In the last paragraph on p. 2021, the authors compare the lower quartile of walk times to the upper quartile of walk times. The adjusted HR for mortality is 3.23, $p<.001$. Relate the confidence interval that is given to the *P*-value. For CVD, $p<.03$. How is this reflected in the confidence interval?

10. Figures 2 and 3 are Kaplan-Meier plots. These plots show how mortality, etc, proceeded as the trial went on. The lower right plot in Figure 2 is for CVD in Women. The plot says "*P*=.16." What is being tested? Why is this hypothesis important? [Hint: Compare the plot in the lower right to the other three plots.]

11. Table 4 on p. 2025 compares cardiovascular response to the various outcomes (mortality, etc.) This table is discussed in the last paragraph on p. 2022. They say, in part, "Higher heart rate response and faster heart rate recovery were both inversely associated with mortality ..." What is it about Table 4 that means "inversely"? What can you say about *P*-values?

12. In further discussing Table 4, the authors say "these associations were largely explained by health conditions ... and faster ... walk times." How does this show up in Table 4? What is the significance of "and" in their statement? That is, is the effect largely explained by health conditions *without* using walk times?